S0-AKT-571

THOMISTIC PAPERS
VI

THOMISTIC PAPERS

Thomas A. Russman, O.F.M. Cap.
General Editor

THOMISTIC PAPERS
VI

John F. X. Knasas
Editor

CENTER FOR THOMISTIC STUDIES
University of St. Thomas
3800 Montrose Boulevard
Houston, Texas 77006

LC 83-73623
ISBN 0-268-01886-3 (cloth)
ISBN 0-268-01887-1 (paper)

Manufactured in the United States of America

CONTENTS

EDITOR'S PREFACE

In 1981 the Center for Thomistic Studies published under the editorship of Victor B. Brezik, C.S.B., a volume entitled *One Hundred Years of Thomism: Aeterni Patris and Afterwards, A Symposium.* The occasion of the volume was twofold: a 1979 symposium celebrating the centenary of Pope Leo XIII's encyclical *Aeterni Patris* and the soon-to-be launched graduate program, the Center for Thomistic Studies. It is accurate to say that in the mind of Father Brezik, the Center's founder, the articles contained in *One Hundred Years* (whose authors include: L. Boyle, A. Maurer, D. Gallagher, R. McInerny, M. Crowe, R. Henle, L. Sweeney, J. Owens, V. Bourke) comprised something like a manifesto of the Center's motivating ideals. In that respect *Aeterni Patris* had not become dead letter. Rather, its call to philosophize in the manner of Aquinas for the good of today's world still sounded a harmonious note deep in the hearts of those present at the initiation of the Center. Also, Etienne Gilson and Jacques Maritain stood forth as giants to exemplify how to respond to the wishes of the Pope and as reliable guides to Aquinas' distinctive and perennially true philosophical principles. Especially noteworthy were their expositions of Aquinas' metaphysics of *esse* and analogy and Aquinas' epistemology of direct realism.

Ten years after the *Aeterni Patris* symposium, Gerald McCool, S.J., published a narrative on the neo-Thomist period entitled *From Unity to Pluralism: the Internal Evolution of Thomism.* In a dazzling display of historical erudition, McCool sought to drive home the following theses. First, the neo-Thomism called for by *Aeterni Patris* was deceased. This Thomism presupposed the existence of a single perennially valid philosophy. Ironically killing that conception was the work of key neo-Thomists themselves. Joseph Maréchal especially, but later Karl Rahner and Bernard Lonergan, worked out interpretations of Aquinas in which empirically derived concepts were always limited by historical conditions. What remained stable and invariant was a subjective factor: the mind's dynamism to the infinite. By an appeal to a Kantian-style transcendental method--something claimed to be not foreign to the works of Aquinas, one could explicate this factor and provide the necessary underpinnings for Catholic theology.

Second, despite his strong on-the-record stance against a Thomistic legitimacy for the modern critical problem, Etienne Gilson could be enlisted in the above Transcendental Thomist project as it undermined the dream of *Aeterni Patris*. Gilson's stand on Aquinas' "Christian philosophy" seemed to wed philosophy to theology in a way counter to the more autonomous and rational philosophy called for by *Aeterni Patris*. Also, Gilson's own historical labors in the medieval period proved that no single perennially true philosophy constituted Scholasticism. Everywhere there was difference. In fact this difference extended into the great Renaissance Thomists, Cajetan and John of St. Thomas.

Hence, third, despite his profound philosophical genius, Jacques Maritain, who drew heavily from Cajetan and John of St. Thomas, was unable to transcend a limited and historically conditioned form of Thomism.

I believe that enough has been said to show that apart from any claims to know the intentions of its author, logically speaking *From Unity to Pluralism* stands opposed to the motivating ideals permeating *One Hundred Years of Thomism*. Out of this opposition and the continued presence of the Center for Thomistic Studies emerged the idea of *Thomistic Papers VI*. By enlisting as many as possible of the previous contributors to *One Hundred Years*, as well as others, volume six would devote itself to assessments of McCool's narrative.

As the present work attests, that idea has now been realized. As editor I will not try to preview here the many varied points made in the articles. The contributors all ably speak for themselves. It is, however, accurate to say that the predominant judgment is that McCool's theses ought to be contested. Plenty of fight seems to remain in today's Thomists. As editor of a volume that has taken a contentious turn, I sincerely wish to conclude this preface with the following points.

I know that I speak for all contributors when I say that among us is a profound respect and genuine admiration for the corner Father McCool has carved out for himself as a narrator of the neo-Thomist heritage. Unfortunately, that respect is not conveyed by agreement with him. It is conveyed, hopefully, by the seriousness with which his theses are taken and by the honest attempt to understand his reasons for them.

No one is out to un-Thomisticize Father McCool. For me a Thomist is this: a philosopher whose seminal ideas derive from the texts of Aquinas as that philosopher understands them. That suffices to establish the Thomistic fraternity just as love of wisdom, as distinct from actual possession, founds the philosophic fraternity. Certainly in that sense of "Thomist," Father McCool is a Thomist and an eminent one at that. But Thomists also need an understanding of themselves that allows them to call each other wrong when that is necessary.

Finally, as the history of philosophy shows, public conflict in the form of clear, incisive, and accurate criticism is not the death of philosophy but its life-blood. One should not regard it as an embarrassment to the parties involved nor should it be stifled by calls to present a united front for political reasons. *From Unity to Pluralism* is an immensely faceted presentation and one could always find more to say about it. It is my best judgment, however, that the contributions to *Thomistic Papers VI* achieve that level of constructive criticism. I would enjoy nothing more than to shatter the silence that has fallen between the great Thomistic camps. Catholic philosophers might well leave this fascinating century, so engagingly told by Father McCool, with another Thomist renaissance.

Thomistic Papers VI is pleased to include in an appendix the Center for Thomistic Studies 1993 Aquinas Lecture by Dr. Mary Rousseau. The lecture is entitled "Thomistic Personalism and Today's Families" and was presented in January at the University of St. Thomas.

MARITAIN AND GILSON ON THE QUESTION OF A LIVING THOMISM

Victor B. Brezik, C.S.B.

On initial analysis, the phrase "Living Thomism" presents not one problem but two. How is one to understand the term "living" in this context? Does this imply some sort of continuity of doctrine as well as new additions, new applications to changing historical and cultural developments? The term "Thomism" is no less clear. Does the term refer to the writings of St. Thomas himself or only to a body of doctrine inspired by the writings of St. Thomas and held by those who profess to follow the teaching of St. Thomas? Add to this the question whether "Thomism" designates a philosophy standing apart from theology or simply the theology of St. Thomas together with its philosophical content?

I am going to explore these questions with particular reference to the views of Maritain and Gilson. Does "Living Thomism" have a common meaning for them both? This question is provoked by the charge that Gilson repudiates the Thomist tradition whereas Maritain drew his understanding of St. Thomas from the classical Thomist commentators. In his book *From Unity to Pluralism*,[1] Father Gerald McCool, S.J., distinguishes their teaching in terms of a systematic pluralism within Thomism, as though their approaches to the doctrine of St. Thomas divides them as representatives of opposed versions or interpretations of the teaching of Aquinas. I propose to consider their views with respect to the question of a living Thomism. The expression would seem to bear special significance for those who regard the teaching of the Angelic Doctor as having contemporary value beyond mere historical interest. I

---

[1] Gerald A. McCool, S.J., *From Unity to Pluralism: The Internal Evolution of Thomism* (New York: Fordham University Press, 1992).

have made some references to myself simply to give the question a more familiar setting.

I

At the time I first enrolled as a student at the University of Toronto, Etienne Gilson was already the Director of Studies at the fledgling Institute of Mediaeval Studies founded in 1929 at St. Michael's College. Jacques Maritain came to lecture there at the end of 1932 at the recommendation of Gilson. The two philosophers were recognized even then as two outstanding Thomists of this century.

I take the word "Thomist" to refer to those who profess to follow the teaching of St. Thomas Aquinas. Accordingly, I would not call St. Thomas himself a Thomist. His teaching is the doctrinal source of the school of Thomists. In this respect, however, as the original deposit of doctrine, the teaching of St. Thomas is usually included under the broad name of Thomism.

During the 1930's and 1940's, while I was a student and later a young teacher in Toronto, the faculty as well as the students were clearly perceptive of the different approaches and presentations of the doctrine of St. Thomas Aquinas in the lectures of these two Thomist professors. Each in his own characteristic way and style drew his teaching principally from the texts of the Angelic Doctor, astounding their listeners with the relevance and vitality this thirteenth-century teaching still possessed. It was evident that a doctrine that could spark the intellectual energy of two creative minds like those of Gilson and Maritain gave witness thereby to its own intelligible richness.

Differences between the Thomism of Maritain and the Thomism of Gilson were accounted for by some as differences between Gilson the historian and Maritain the metaphysician. At that date, Gilson was widely recognized for his outstanding contributions to the history of medieval philosophy. His metaphysical interests were not yet generally known. Maritain, on the other hand, early in his career was identified as a speculative thinker rather than as an historian.

Thus they represented two distinct approaches to the thought of St. Thomas Aquinas.[2] One of these has been called "the historical-textual approach." It consists of reading carefully the works of St. Thomas in the best critical editions, studying his teaching in its historical setting, and endeavoring to determine its precise meaning. This describes more or less accurately the method of approach taken by Gilson. The second approach to the thought of Aquinas consists in re-thinking this rich body of doctrine in relation to constantly changing contemporary problems and thought-patterns and their solutions. Maritain seems to have adopted this second approach to a large extent. His writings and lectures were consistently addressed to working out present problems in the light of Thomistic thought.

Both of these approaches are indispensable for giving the teaching of St. Thomas a continuing value. For how could one re-think Aquinas' thought faithfully and authentically without the discipline of a prior historical-textual study? At the same time, how could one avoid reducing his doctrine to a museum-piece, without re-thinking it with reference to the vicissitudes of history and to cultural changes?

Despite the dissimilitude in their presentation of Thomism, emphasized not without some justice by Father McCool, these two Thomists were in accord in subscribing to the notion of a "Living Thomism" as they were in accord in their general understanding of the essential nature of Thomism. No more than Maritain, did Gilson regard the teaching of St. Thomas as dated by

---

[2] See Vernon J. Bourke, "The New Center and The Intellectualism of St. Thomas," in *One Hundred Years of Thomism*, ed. Victor B. Brezik, C.S.B. (Houston: University of St. Thomas), 1981, p. 169.

the thirteenth century. The whole effort of their teaching and writing careers was clearly aimed at making St. Thomas come alive for their contemporaries in the twentieth century.

II

This aim, on Maritain's part, already evident from his writings prior to his first visit to this continent, was reaffirmed during this initial visit to Toronto. Maritain was fifty years of age at this time. As a freshman at St. Michael's College, I was privileged to attend his very first lecture in Toronto, given in the old study hall on Bay Street, the largest classroom at the College at that time. I still have in my possession a mimeographed copy of the English translation of that first lecture. In presenting his views, I intend to draw upon this lecture.

In his introductory remarks, Maritain explained his reason for braving the storm at sea to come to America. Europe, he declared, had years ago brought the Christian faith to these shores. The time was approaching, he thought, when America would have to bring the faith back to Europe. The thinking of the new would, he said, will have a decisive influence on the future and on the culture of the mind. That is why he crossed the ocean, for the philosophy of St. Thomas offers salvation for the intelligence.

Shortly before his voyage to America, Maritain had written his little treatise *Sept leçons sur l'être*, translated later into English under the title: *A Preface to Metaphysics*.[3] The substance of this first lecture in Toronto was taken from the First Lecture of this small

---

[3] Jacques Maritain, *A Preface to Metaphysics: Seven Lectures on Being* (New York: Sheed and Ward, 1948). Reprinted as A Mentor Omega Book, (New York: The New American Library of World Literature, 1962).

book.[4] In his Toronto lecture, Maritain proclaimed his support for "a living Thomism, not an archaeological Thomism." We must, of course, study Thomism historically, he said, in order to know what it was and what it is now. Yet we must think of it not only historically but in connection with contemporary problems.

For this reason, Thomists, he observed, have a twofold obligation to fulfill. First, to defend the stability and permanence of traditional thought against the individualism of modern thought and the mistaken conception of progress, which places undue confidence in novelty simply because it is novelty. The second Thomist obligation is to defend the vitality and development of traditional thought against the immobilism and rigidity of some scholastics. The wisdom of St. Thomas, Maritain affirmed, is always young. It renews its youth, constantly renovates and rejuvenates itself. It is a growing body. In short, Thomism must be both traditionalist and progressive, innovative.

Maritain distinguishes between two types of progress. The first is by substitution or successive replacement and corresponds to progress in science[5] where the "problem aspect" predominates. The second is progress by deepening insight and corresponds to

---

[4] In a subsequent work, Maritain recommended that the First Lecture of *Sept leçons sur L'être* (*A Preface To Metaphysics*) and *The Degrees of Knowledge* be consulted "concerning the way in which I think we ought to conceive the effective progress of philosophy." Jacques Maritain, *Science and Wisdom*, trans. by Bernard Wall (New York: Scribner's, 1940), p. 81, note 1.

[5] See an exemplification of the progress of science by substitution in Gilson's address to the 1952 Annual Meeting of The American Catholic Philosophical Association entitled: "Science, Philosophy, and Religious Wisdom," in Anton C. Pegis, ed., *A Gilson Reader* (Garden City, New York: Doubleday Image Book, 1957), pp. 213-216.

wisdom where the "mystery aspect" predominates. This second type is meant to define and locate the kind of progress he envisioned for Thomism.

The work facing modern Thomists, according to Maritain, is similar to the task which confronted St. Thomas in the thirteenth century. St. Thomas had to purify the stream of Christian thought coming down from St. Augustine, scrape off, as it were, the rust of foreign accretions, so that it might flow on with pure waters. It was not easy. Rigid traditionalists stood in the way, charging him with "novelties." Gilson, Maritain remarked in his lecture, speaks of "hard-boiled theologians." Similar obstacles exist today. Thomists must overcome contemporary philosophical "stand-patters," Maritain urged, who swear by their textbook learning and oppose all "innovation," that is, they oppose honest and enlightened efforts at improvement and development.

The growth of Thomism, for Maritain, must be organic, resembling that of a child which, growing into the adult state, undergoes change and develops, yet remains the same person throughout. The future here grows out of the past and is filled with all that the past has found good and true. "Real development," Chesterton wrote (quoted by Maritain in his lecture), "is not leaving things behind, as on a road, but drawing life from them as from a root."[6] As an organism, Thomism can assimilate, as every living organism assimilates, the material of its environment. This entails full contact of Thomists with non-Thomists. Such progress, Maritain believed, does not involve change of substance but of mode. "Old truths in new dress, new forms of presentation, new perspectives on the same landscape." As Maritain conceived the changing *modes* of philosophizing, the concepts themselves do not change. The same, identical concept may be rediscovered by a

---

[6] Gilbert Keith Chesterton, *The Victorian Age in Literature* (London: Oxford University Press, 1961), Chap. 1, p. 10.

different *mode*, a different approach, from a different angle or point of view. Such differences of *mode* inevitably involve differences of expression, language, vocabulary.

III

What was Gilson's reaction to these views of Maritain? To my knowledge, there was no public response. In successive visits to Toronto, Gilson carried on his lectures in his usual manner. On occasions, during these years, when Maritain and Gilson were in Toronto at the same time, their relations were obviously cordial and their friendship warm. One could even detect in them a kind of mutual admiration. On one occasion, when addressing a group on campus, Gilson remarked that it would take a hundred years to measure the achievement of Jacques Maritain. At times when they jointly spoke to informal gatherings of faculty and students, Maritain was always deferential to Gilson.

Much later, in 1964, Gilson's little book entitled *The Spirit of Thomism* appeared.[7] It is of interest to note that Chapter IV of this book bears as its title the very phrase used by Maritain in his first lecture in Toronto, namely, "A Living Thomism."

In this chapter on a living Thomism, Gilson considered what there is to expect regarding the future of Thomism. He discerned a number of obstacles on its road to success. The primary obstacle is "its religious inspiration." This obstacle is so formidable as to have prompted some Thomists to attempt a complete separation of Thomistic philosophy from Thomistic theology. While it is true that St. Thomas himself established a clear-cut distinction between reason and faith, philosophy and theology, he did not keep them

---

[7] Etienne Gilson, *The Spirit of Thomism* (New York: P. J. Kennedy, 1964). Reprinted, New York: Harper and Row, Harper Torchbooks, 1966.

apart. How is one to overcome this obstacle and satisfy such rationalist-minded Thomists? Gilson offers a simple answer: teach Thomism just as it is. To remain faithful to Thomism, he claims, there is no other choice. We must observe the advice of Thomas Aquinas: follow the guidance of natural reason and leave the rest to God.

A more insurmountable obstacle, Gilson thinks, is a purely philosophical one. In an age in which systems of idealism prevail, many philosophers are likely to be turned against Thomism because it professes realism. The philosophy of St. Thomas is a philosophy of being, understood in terms of "the act in virtue of which being is, and is being, namely that of *esse*."[8] Although it leads to no system, this principle inserts light and order into the manifold of reality. The first thing necessary, then, to rejuvenate Thomism is to restore the right interpretation of the first principle. Immediately following from this first characteristic of a living Thomism is a second: since in beings *to be* comes first and to be is an act, the real world is "not made up of static essences but of acting, operating and causing beings."[9]

The third and most serious obstacle to the survival of Thomism is, according to Gilson, the actual failure of most Thomists to provide proofs of its continuing vitality. Much has been done, he admits, to keep Thomism alive, the best thing, perhaps, being the effort of many to clarify the first principles of Thomism. But the knowledge of first principles, although its higher part, is not the whole of a philosophy. The brunt of the objection is that scholasticism has long tended to be but a more or less repetitious kind of school teaching, short on producing new ideas and backward in explaining rationally the modern world. Putting it bluntly, the real weakness of modern scholasticism has been its

---

[8] *The Spirit of Thomism*, Harper Torchbooks, p. 88.
[9] *Ibid.*, p. 89.

"sterility." It is not the principles themselves that are to blame. They are perfectly sound. The blame must be placed on the neglect of Thomists to put the principles to good use. As St. Thomas himself so successfully applied sound principles to the world in which he lived, Thomists today have only to turn their light on the world of things around them to find plenty to see and say. Briefly, Thomism needs to be creative, recognizing, however, that the present world: scientific, political, social and economical, has ceased to be the world familiar to St. Thomas. Accordingly, a living Thomism, says Gilson, should in the light of the permanently valid principles of the Thomistic metaphysics of being, devote itself to the urgent task of criticizing, interpreting and ordering the mountain of material that has been piling up since the time of St. Thomas.

As I read this, what Gilson seems to be proposing is that it is not enough merely to learn philosophy; it is necessary actively to philosophize and to do so in the context of the present world. Modern Thomists must be contemporary in their thinking after the manner in which St. Thomas, using principles drawn from Aristotle, was contemporary in his day.

I do not recognize anything in these views of Gilson, if properly understood, that is in open discord with the stated views of Maritain. In fact, as an example among Thomists who had already demonstrated the creative possibilities that lie open to Thomism, Gilson singled out his colleague, Jacques Maritain.

> In the fields of the philosophy of nature, of political economy and the so-called 'human sciences,' the example of Jacques Maritain clearly shows how it is still possible today to renovate ancient concepts and to open new fields of investigation. The philosophy of art, illustrated by the same philosopher, clearly shows that in certain cases Thomism is bound to create if it is still to live. For

> indeed Thomas himself has said precious little, if anything, about the fine arts.[10]

Incidentally, Father McCool states that Gilson "saw little value in Maritain's Thomistic philosophy of nature."[11] Can this statement be fully reconciled with the above quotation?

In a note to this same Chapter IV in *The Spirit of Thomism* (note 10, p. 124), Gilson adds:

> Jacques Maritain's contribution to the development of a moral philosophy is authentically Thomist in its inspiration and yet resolutely modern in its way of handling problems. It represents a deeply original part of his work.

In connection with Gilson's approval here of Maritain's style of exemplifying a living Thomism, one should not forget Gilson's ecomium of Maritain in Gilson's *The Philosopher and Theology*.[12] There he refers to Maritain as "the only Thomist in contemporary France whose thought was lofty, bold, and creative, capable of meeting the most urgent problems." Gilson adds: "It is not necessary to read many pages from Jacques Maritain to realize that one is dealing with one of the best French writers of our time."

---

[10] *Ibid.*, p. 100.

[11] *From Unity to Pluralism*, p. 194.

[12] Etienne Gilson, *The Philosopher and Theology*, trans. by Cecile Gilson (New York: Random House, 1962), pp. 201, 202. In his survey elsewhere of French and Italian philosophy, Gilson declared that "French neo-Thomism bore its most precious fruit in the work of Jacques Maritain." Etienne Gilson, Thomas Langan, Armand A. Maurer, C.S.B., *Recent Philosophy Hegel to the Present* (New York: Random House, 1962), p. 352. On Gilson's account of Maritain, see pp. 352-354 and note 50, p. 789.

For Gilson a right understanding of the meaning of *being* is the very key to Thomistic metaphysics.[13] On this critical point, in his fifth edition of *Le Thomisme*, Gilson, after a series of quotations from Maritain's *A Preface to Metaphysics*, concludes with the following words of endorsement: "It could hardly be put better."[14] In the same work (p. 463, note 37), Gilson refers to Maritain as "one of St. Thomas's most profound interpreters."

These tributes of Gilson to Maritain as a fellow Thomist do not read like the remarks one would make of an opponent whose conception of Christian philosophy, one thinks, "could not be," in the words of Father McCool, "an authentic interpretation of St. Thomas' own thought."[15] Not only are these tributes an endorsement of the living Thomism of one whose thought is "lofty, bold, and creative"; they also indicate a congeniality of purpose between the two Thomist philosophers.

IV

Yet one must not rush hastily to conclusions. There are enough points of difference to cause one to question whether the *Living Thomism* of Gilson and the *Living Thomism* of Maritain are in full accord. As to the need for an historical study of Thomism, the revival of a correct interpretation of metaphysical principles, the affirmation of St. Thomas' realism, the stress on the primacy of *esse* in the Thomist concept of being, the

---

[13] Etienne Gilson, "What Is Christian Philosophy?" in *A Gilson Reader*, p. 190. Cf. Etienne Gilson, *The Christian Philosophy of St. Thomas Aquinas*, trans. by L. K. Shook, C.S.B. (New York: Random House, 1956), p. vii: "the notion of the act of being (*esse*) . . . is the very core of the Thomistic interpretation of reality."

[14] *The Christian Philosophy of St. Thomas Aquinas*, p. 365.

[15] *From Unity to Pluralism*, p. 194.

urgency for innovation and creativity, the necessity of contact with contemporary thought, the recognition of a certain neglect among Thomists, these are points on which their views of a living Thomism agree or at least can be adjusted. But to harmonize their views on questions such as the relations of philosophy and theology, the autonomy of philosophy, the role of a Thomistic tradition and the way to teach Thomism may not be as easy.

Both Maritain and Gilson firmly adhered to the concept of Christian Philosophy and stood together in its defense during the controversy that developed around it and which in a subdued manner survives to this day. Much has been written on this question of Christian Philosophy, which excuses me from going into detail. My interest here is only to note that in the midst of the debate, Maritain and Gilson remained partners on the same side. Gilson thought he had adequately established the fact of Christian Philosophy as well as exemplified its variations in his Gifford Lectures at the University of Aberdeen, published in English under the title of *The Spirit of Mediaeval Philosophy.*[16]

Already at that time in 1931 and 1932, Gilson pointed out that "the concept does not correspond to any simple essence susceptible of abstract definition." It corresponds "much rather to a concrete historical reality as something calling for description." Such a philosophy, he said, while keeping the two orders formally distinct, nevertheless regards Christian revelation "as an indispensable auxiliary to reason."[17]

Maritain, in his turn, explained the nature of Christian philosophy by distinguishing philosophy, considered abstractly in its essence and as specified by an object naturally knowable to reason, from philosophy

---

[16] Etienne Gilson, *The Spirit of Mediaeval Philosophy* (Gifford Lectures 1931-1932), trans. by A. H. C. Downes (New York: Scribner's, 1936).

[17] *Ibid.*, p. 37 (All three quotations).

taken concretely and considered in a certain state as it exists in the human soul, "under conditions of performance, of existence and of life."[18] Philosophy, he contended, is Christian only in the latter sense, considered in the order of exercise, not in the order of specification (except practical philosophy), that is, not as an essence.

When Gilson says[19] that scholastic philosophy is not distinguished from other philosophies by its essence but rather as the best way of philosophizing, he is invoking much the same distinction between the order of specification and the order of exercise made by Maritain. What is this best way of philosophizing? It is the way recommended by Pope Leo XIII in his Encyclical *Aeterni Patris*, namely, the way of philosophizing of those "who to the study of philosophy unite obedience to the Christian faith."[20] The result of such philosophizing turns out a product with identifiable differences from non-Christian philosophies, yet at the same time differing individually among those who use this method of philosophizing. Gilson made this clear in his

---

[18] *Science and Wisdom*, p. 81. For Maritain's views on Christian Philosophy, see also his *An Essay on Christian Philosophy*, trans. by Edward H. Flannery (New York: Philosophical Library, 1955). In the second volume of the French edition (p. 290) of *The Spirit of Mediaeval Philosophy*, in reference to Maritain's *An Essay on Christian Philosophy*, Gilson wrote: "I may say, then, that Christian philosophy is an objectively observable reality for history alone, and that its existence is positively verifiable by history alone, but that once its existence has been thus established its notion may be analyzed in itself. This ought to be done as Mr. J. Maritain has done it; I am in fact in complete agreement with him." Quoted from *An Essay on Christian Philosophy*, p. xi.

[19] Etienne Gilson, "Historical Research and the Future of Scholasticism," *The Modern Schoolman*, 29 (1957), 1-10. Reprinted in *A Gilson Reader*, p. 165.

[20] "What Is Christian Philosophy?" in *A Gilson Reader*, p. 186.

individual studies of the Christian philosophies of St. Augustine, St. Bonaventure, St. Thomas Aquinas and Duns Scotus.

Gilson and Maritain heartily concurred with Pope Leo's Encyclical that "among the Scholastic Doctors, the chief and master of all, towers Thomas Aquinas."[21] Like Thomas, their common master in philosophy, Gilson and Maritain philosophized keeping their minds turned toward their Christian faith. That this way of philosophizing confuses philosophy with theology, they mutually disclaimed. At the same time, they equally rejected any complete separation of philosophy and theology. It is, nevertheless, precisely with regard to the relation of philosophy to theology that differences in their views of a living Thomism begin to show.

Gilson's historical studies of medieval thought led him to two conclusions about philosophy in the Middle Ages. The first is that "research in medieval thought, which began by being concerned with the philosophies of the middle ages, is tending more and more to restore these philosophies within the theologies which contain them"[22] A second conclusion taught by this historical research is that "the more we integrate the philosophies of the middle ages within their theological context, the more their originality becomes apparent."[23] Gilson conceives other ways of expressing this conclusion. For example, "it is while serving theology that philosophical thought became creative." Again, "the more a master was a great theologian, the more he was a great philosopher." In short, it is precisely to its role as

---

[21] Etienne Gilson, ed., *The Church Speaks to the Modern World, The Social Teachings of Leo XIII* (Garden City, New York: Doubleday Image Books, 1954), p. 43. Also Jacques Maritain, *St. Thomas Aquinas, Angel of the Schools*, trans. by J. F. Scanlan (London: Sheed and Ward, 1942), p. 204.

[22] *A Gilson Reader*, p. 159, 160.

[23] *Ibid.*, p. 161.

an instrument of theology that medieval philosophy owes its "fecundity."[24]

This second conclusion, in Gilson's view, applies to the philosophy of St. Thomas Aquinas as much, if not even more, as it applies to the philosophies of other medieval theologians. It is wrong to think that Aquinas, or any other medieval theologian for that matter, founded his theology on any philosophy, even the philosophy of Aristotle.[25] What he did was to make use of philosophy within the light of faith with the result that philosophy came forth transformed. The metaphysics of St. Thomas is his own metaphysics and to identify his metaphysics with the Aristotelian metaphysics of being is to have an improper understanding of it. His metaphysics is a new metaphysics, which shares in the permanence of the light of faith within which it was born.[26]

What is most important about a metaphysics, Gilson thinks, is its conception of the first principles. This is precisely what gives the metaphysics of St. Thomas the right to be the doctrinal norm in Christian philosophy. His metaphysics is based on a conception of these principles that not only agrees perfectly with Sacred Scripture but at the same time "assigns to metaphysics the deepest interpretation of the notion of being ever offered by any philosophy."[27] A strong statement indeed. Does this mean that for Gilson the history of metaphysics stops at St. Thomas Aquinas? In other words, can there be any further deepening of the notion of being than that found in the doctrine of St. Thomas? Such a possibility is not entirely ruled out.

---

[24] *Ibid.*, p. 162.
[25] *Ibid.*, p. 163.
[26] *Ibid.*, p. 164.
[27] *The Philosopher and Theology*, p. 234.

Gilson himself simply does not pretend to see that far into the future.[28]

What, then, can be said about the progress of Thomism as a Christian philosophy? Gilson allows that its progress may be unending, on the condition that it remains faithful to the truth of its principles. New discoveries made in any order of knowledge will provide occasions for its fecundity to show itself. "Every progress whatever can be for it an occasion of progress."[29] Long reflection on the truth of St. Thomas' metaphysics made Gilson see it as a light capable of absorbing every other light. The Thomistic notion of *esse* (to be), he claimed, is ultimate by its very nature. "It lays the foundation of metaphysical knowledge for all time."[30]

Would Maritain himself concur with these views of Gilson? In his *A Preface to Metaphysics*,[31] writing of the relation between philosophy and theology, Maritain pointed out that, since it is based on the *Word* of God, theology must obviously be permanent. As a science rooted in the faith, theology does develop and progress, not however by successive substitutions, rather by more intimate penetration into its subject matter. What is more, in its development, theology uses philosophy as a means and instrument, with the result that philosophy, too, in its own fashion must also be permanent.

This permanence, Maritain thinks, has to be considered by Thomists in their work of renewing the doctrine of St. Thomas. The work must be carried on "without detriment to the fixity of principles," since the Thomistic philosophy is "securely based on true principles."[32] When Gilson described Christian

---

[28] *Ibid.*, p. 235.

[29] *Ibid.*, p. 235.

[30] *Ibid.*, p. 236.

[31] *A Preface to Metaphysics*, p. 17.

[32] *Ibid.*, p. 19.

philosophy as "the unfolding of a progress from a truth itself not susceptible of progress,"[33] was he saying something out for harmony with Maritain?

V

Maritain and Gilson both trace the attempt to separate philosophy from theology to the Averroist movement active at the time of St. Thomas.[34] Its revival in the sixteenth century turned out to be a preparation for a more thorough seventeenth century revolution during which Descartes separated philosophical wisdom from theological wisdom. Henceforth, philosophy was no longer to play a ministerial role with respect to theology. It now declared its full autonomy as a separated discipline.

How did Gilson and Maritain respond to this historical development? It is here especially that there appears to be some parting of the ways. Insofar as the Cartesian revolution constituted an effort to effect a complete break between philosophy and theology, Gilson and Maritain were equally and staunchly opposed to it. Neither of them envisioned Thomism as a "separated philosophy"[35] or a "pure philosophy"[36] where philosophy and faith would be kept entirely apart in the Cartesian manner. They both agreed that St. Thomas did not develop philosophy as a distinct

---

[33] *The Philosopher and Theology*, p. 233.

[34] *Science and Wisdom*, p. 28; *A Gilson Reader*, p. 157. Gilson states that "philosophism lies at the bottom of Averroism, as indeed it lies at the bottom of the positions of Averroes himself." *History of Christian Philosophy in the Middle Ages* (New York: Random House, 1955), p. 408. See also his *Reason and Revelation in the Middle Ages* (New York: Scribner's, 1938), Chap. 2.

[35] *Science and Wisdom*, p. 82.

[36] *History Of Christian Philosophy In The Middle Ages*, p. 542; *A Gilson Reader*, p. 172.

discipline standing outside theology. He developed it only within his theology and as an instrument of that theology. The difference is that Gilson defended this inseparable state of Thomistic philosophy as a permanent condition, since that is the state in which St. Thomas left it,[37] whereas Maritain did not. Maritain did not conceive its position of subservience to theology among medieval theologians as the normal and natural state for philosophy. In this respect, he viewed the autonomy of the sciences and of philosophy in post-Cartesian times as a "precious gain."[38]

---

[37] To the question: "In what sense may one speak of a philosophy of St. Thomas," Gilson offers two possibilities. First, Thomistic philosophy may be taken as a complete exposition of the philosophical notions found in the works of St. Thomas, including materials borrowed from his predecessors, and elaborated into a doctrinal synthesis. There have been instances of such expositions of Thomistic philosophy. Secondly, Thomistic philosophy may refer to a synthesis of notions present in St. Thomas' collected works which are truly his own notions, distinct from those of his predecessors. Gilson's interest was to show in what such an original Thomistic philosophy consists. And since, as Gilson thought, the most original aspects of St. Thomas' philosophy are in general lodged within his theological works, it would seem natural in setting forth his philosophy to observe the order of his theology. To extract these philosophical notions from their theological setting in an attempt to reconstruct them according to some sort of philosophical order would be to presume that perhaps St. Thomas himself meant to set up a philosophy with purely philosophical ends. Not only is there risk of altering the nature of his philosophy in such an attempt to set it free of its theological locus; the attempt is not in the least necessary. According to Gilson, it is not at all impossible to present the philosophy of St. Thomas in the order of his theology without thereby confusing reason with faith. The fact is that St. Thomas himself has done it and the difficulties of doing it after him disappear with a clear understanding of how he understood theology. See: *The Christian Philosophy of St. Thomas Aquinas*, pp. 7-9.

[38] *Science and Wisdom*, p. 33.

Gilson could not quite see it this way. To be a Thomist for Gilson and to learn Thomism, it is necessary to read the theology of St. Thomas and in doing so, imbibe the philosophy which it employs and contains. So intimate is the relationship of this philosophy to the theology which gave it birth and to which it owes its originality, that the deep meaning of its thought as entirely engaged in the service of the faith, will never be fully understood by someone who does not possess the Christian faith.[39] In fact, Gilson says, "the most original notions and the deepest, in the doctrine of Saint Thomas reveal themselves only to him who reads it as a theologian."[40] Since this is the case, Gilson proposes

---

[39] *The Philosopher and Theology*, p. 210.

[40] *Ibid.*, p. 211. In an effort to clarify the nature of the philosophy of St. Thomas, Gilson uses the distinction St. Thomas makes between the *revelatum* and the *revelabilia* (*ST*, I, 1, 3). "The revealed," in Gilson's interpretation, refers to "all knowledge about God beyond the grasp of human reason." (*The Christian Philosophy of St. Thomas Aquinas*, p. 11). It comprises all that knowledge which human beings can acquire only through divine revelation (*Ibid.* p. 12). Besides the essentially revealed truths, i.e., the *revelatum*, revelation itself includes many other truths which are accessible to human reason without the aid of revelation. This is the proper area of "the revealable," which contains the philosophical elements St. Thomas employed in his theology. These philosophical elements which the theologian uses remain truly philosophical in their essence and nature even though the theologian views them in a higher light for theological ends, "as a possible help in the great work of man's salvation." (See Etienne Gilson, *Elements of Christian Philosophy* [Garden City, New York: Doubleday, 1960], pp. 34, 35. Reprinted as A Mentor-Omega Book [New York: New American Library of World Literature, 1963,] pp. 36, 37). This precisely is the perspective in which Gilson wishes to examine the philosophy of St. Thomas, since it is the perspective of St. Thomas himself (*The Christian Philosophy of St. Thomas Aquinas*, p. 15). It can be examined, he acknowledges, from other points of view. For instance, some profess to reconstruct St. Thomas' teaching in the philosophical

some interesting, though unusual, suggestions regarding the teaching of Thomist philosophy to Catholic students. "If you want to teach your students both metaphysics and ethics, teach them straight theology."[41] Why deceive oneself in thinking that by taking theology out of the *Summa* and putting in another book, one is setting things in order by thus sorting out what is philosophy and what is theology as if the *Summa* were a mixture of both? The result can only be an adulterated philosophy, on the one hand, and an adulterated theology, on the other. The truth is that the *Summa* is not a mixture of philosophy and theology; rather, by being put at the service of faith in the *Summa*, the water of philosophy is changed into the wine of theology.[42] A reasonable corollary to teaching philosophy through the study of theology would seem to be: all professors of philosophy should be theologians. Gilson cites without proposing as his own this recommendation of the Jesuit *Ratio*

---

order going from things to God rather than following the theological order which proceeds from God to things. Gilson notes two dangers in such a procedure. The first is to end up substituting the philosophy of Aristotle for that of St. Thomas; the other is the contemporary error of flatly contradicting the philosophy one pretends to teach. Gilson looks upon the attempts to isolate St. Thomas' philosophy from his theology as having the result of presenting his philosophy in the Cartesian manner whereby everything is "considered by natural reason without the light of faith" (*Ibid.*, p. 442, note 33).

[41]"Thomas Aquinas and Our Colleagues," Aquinas Lecture at the Aquinas Foundation at Princeton University, March 7, 1953. Printed in *A Gilson Reader*, p. 292.

[42] *Gilson Reader*, pp. 293, 294. Cf. St. Thomas Aquinas, *In Boethii de Trinitate*, q. 2, a. 3, ad 5; *St. Thomas Aquinas, Faith, Reason and Theology*, Questions I-IV of his commentary on the *De Trinitate* of Boethius, translated with Introduction and Notes by Armand Maurer (Toronto: Pontifical Institute of Mediaeval Studies, 1987), p. 50.

Studiorum of 1586.[43] I doubt not that in actual fact Gilson himself could well enough have met such a qualification for teaching philosophy, as Maritain could have also, even though Gilson referred to himself as a philosopher[44] and Maritain consistently protested that he was treating things only as a philosopher.

One last observation about Gilson's views on the relation of philosophy to theology is that the philosophy of St. Thomas is best expounded by following the theological method according to which St. Thomas himself formulated it. Gilson practiced this method more or less closely himself in his six editions of *Le Thomisme* and the smaller volume entitled *The Elements of Christian Philosophy.* His many other writings in which he expounded Thomism do not observe this same method, which indicates that Gilson did not propose the theological order as the exclusive way of presenting Thomism.

VI

Maritain was in agreement with Gilson that the philosophy of St. Thomas played a ministerial and instrumental role in the theological works of St. Thomas and was integrated with his theology.[45] Nevertheless, one of St. Thomas' principal objectives, Maritain claimed, was not only to distinguish philosophy from theology but also thereby to establish the autonomy of philosophy.[46] St. Thomas actually achieved this objective in principle, Maritain says, distinguishing the

---

[43] *A Gilson Reader*, p. 297, note 10.

[44] "The Eminence of Teaching" in *A Gilson Reader*, p. 300.

[45] *Science and Wisdom*, p. 102.

[46] Jacques Maritain, *Existence and the Existent*, trans. by Lewis Galantiere and Gerald B. Phelan (New York: Pantheon, 1948), p. 136. Reprinted, Garden City, New York: Doubleday Image Book, pp. 141, 142.

two disciplines with clarity and firmness. This autonomy, however, has not yet been truly established in fact.[47] Unlike Gilson, Maritain stressed the need to fulfill in point of fact the autonomy fully recognized in point of doctrine.[48]

This is a work cut out for the followers of St. Thomas. In Maritain's opinion, Thomists up to now have not been "very zealous in their effort fully to disengage the proper structure of their philosophy from the methods of approach and the problematics of their theology."[49] Thomist philosophy has too often been presented in the guise of a transposition into the domain proper to reason of a theology deprived of its own light which is faith. Distinct from those of theology, philosophy has its own specific object, its own light and its own mode of dealing with problems and correspondingly its own authentic task. It needs to handle its problems in an autonomous manner in terms of questions arising from experience, not from theology.[50] The same applies to the order of its researches, verifications and judgments. St. Thomas' thought, Maritain admits, is cast in the order proper to theology. Since he wrote no philosophical *Summa*, we cannot say what order he would have followed had he done so.[51] It certainly would have been different from the order of his theology, judging from our present knowledge of what constitutes a philosophical order. The challenge facing present followers of St. Thomas is

---

[47] *Ibid.*, p. 136; Image, p. 142.

[48] *Science and Wisdom*, p. 103.

[49] *Existence and the Existent*, p. 138; Image, p. 143.

[50] *Science and Wisdom*, p.103.

[51] John Wippel suggests that St. Thomas' views regarding the nature, the subject-matter and the method of metaphysics are sufficiently indicated in his writings. See *Metaphysical Themes in Thomas Aquinas* (Washington: The Catholic University of America Press, 1984), p. 32.

to give his philosophy a chance to live outside theology, removed from a theological framework.

It is important not to be misled into confusing Maritain's projections for an autonomous Thomist philosophy with the post-Cartesian rationalist concept of philosophy which demands absolute independence from theology and denies the "infraposition" of philosophy. Philosophy, for Maritain, that is to say, Thomist philosophy, cannot possibly be autonomous in that sense. It can claim only a relative autonomy. A hierarchy of wisdom exists, and rather than denying its subordination, philosophy must strive for self-awareness of its own nature and its special claims as well as its relations to theological and infused wisdom.

In Maritain's perspective, philosophy and culture have suffered great harm from Cartesian "separatism." Nonetheless, the birth of a philosophical or profane wisdom standing on its own feet with its own ends rather than being purely subservient to theology was in accord, he says, with deep historical needs. The differentiation began in the Middle Ages under the aegis of St. Thomas expounding its doctrinal principles was,unfortunately for modern history, "accomplished and realized under the banner of rationalism and division rather than of Christianity and unity." What should have been Christian philosophy, Maritain declared, "became *separated* philosophy. And we have learned our error by bitter and tragic experience." [52]

This emphasis of Maritain on the need to develop the philosophy of St. Thomas as an autonomous discipline appears at first sight to run counter to Gilson's emphasis on studying and learning and teaching this same philosophy within the context of its use by St. Thomas in his theology. Indeed, after examining the work of these two eminent Thomists, it is necessary to ask oneself how far their differences are truly doctrinal

---

[52] *Ibid.*, p. 128.

and how far their differences are accountable to their diverse approaches to Thomism. If one asks whether Maritain proposed a development of the philosophical teaching of St. Thomas in complete separation from his theology, the answer is obviously negative. He always thought of Thomism in terms of a Christian philosophy. Maritain's writings, in fact, are redolent with theological references on almost every page. If one asks whether Gilson claimed that the philosophy of St. Thomas has no legitimate status as philosophy and can exist only as an instrument of theology, the answer, too, is negative. Otherwise, one would be obliged to characterize Gilson's expositions of that philosophy in his books as just so many volumes of theology using philosophy for its own purposes.

VII

Whatever differences exist (and they do exist) between the Thomism of Gilson and the Thomism of Maritain and whether they are differences of emphasis or truly doctrinal differences, they tend to dissolve when considered in the perspective of a Christian philosophy which follows the method of philosophizing proposed by Pope Leo XIII and is exemplified in the works of St. Thomas Aquinas. Although Gilson, especially after experiencing the failure of the Thomist tradition to pass on the authentic meaning of *esse* in St. Thomas' doctrine of being, downgraded dependence on the great Thomist commentators, he did acknowledge a Thomist tradition and considered himself a part of it.[53] Maritain, in turn,

[53] "Life is short and the history of philosophy is growing longer every year. But if any Christian master felt the same indifference with respect to the history of scholasticism, he would be less easily excusable, because this is his own personal history or, at least, that of his own personal philosophical tradition. This tradition is not a dead thing; it is still alive and our times bear witness to its enduring fecundity. There is no reason why this

while he drew inspiration and insight from some of the classical Thomist commentators such as John of St. Thomas and Cajetan, at the same time drank deeply of the original doctrinal springs found in the writings of St. Thomas.[54] Certainly, both Gilson and Maritain would agree that every explanation of Thomistic doctrine must be tested against the very words of the master and that the best interpreter of the writings of St. Thomas are the writings themselves.[55]

Clearly, there are indisputable differences in the presentations of Thomism given by Gilson and Maritain. Father McCool has highlighted some of them in his *From Unity to Pluralism.* How deep a cleavage do these differences make? Do Gilson and Maritain truly

---

fecundity should come to an end." *History of Christian Philosophy in the Middle Ages*, p. 174. Gilson's occasional sweeping criticisms of the Thomist tradition are subject to an overly strict understanding of his intentions. In general, it is true, he favored transcending the tradition to get to the texts themselves of St. Thomas. Nevertheless, he respected the true insights of other Thomists. To cite only one example, in his *The Christian Philosophy of St. Thomas Aquinas* (p. 444, note 1), on the notion of being, he referred favorably to Bañes, O.P., Del Prado, O.P., Oligiati, Forest, Maritain, Pruche, as well as his own work *Being and Some Philosophers*. As to individual points of doctrine which in his eyes did not agree with the texts, he took issue, as, for instance, with Bañes on the *esse* of accidents and with Maritain on several issues. Gilson's numerous citations of and references to other Thomists and their writings, sometimes by way of agreement, sometimes by way of criticism or by way of correction, reveal his vast acquaintance with the Thomist tradition which, far from spurning, he utilized in explaining the doctrine of St. Thomas.

54 Evidence of Maritain's familiarity with the texts of St. Thomas is easily verifiable in his writings.

55 Gilson's rule is: "*Thomas, his own interpreter.* In other words: Do not judge Saint Thomas by his commentators, judge his commentators (including yourself) by Saint Thomas." *The Philosopher and Theology*, p. 207.

represent pluralist interpretations of Thomism as diverse from one another as both their presentations of Thomism are definitely diverse from that of the Maréchal school? Could Gilson, following his own criteria, "hardly deny," as Father McCool states, that neither Maritain's *The Degrees of Knowledge* nor *Science and Wisdom* "are authentic presentations of St. Thomas' philosophy"?[56] Was Maritain's Christian philosophy a perpetuation of the "separated" Thomistic philosophy created in the seventeenth century?[57] These questions taken simply as questions suggest strong differences. Admittedly, the division that Father McCool draws between Maritain and Gilson may serve well the thesis of his book; it nevertheless leaves one wondering in terms of their mutual commitment to a living Thomism whether McCool's division between them is not overdrawn.

When Father McCool goes on to extend the division beyond Maritain to a division between Gilson and other contemporary Thomists,[58] ascribing to their unacceptance of Gilson's conclusions the reason for Gilson's work "seriously undermining the Neo-Thomistic movement,"[59] one feels challenged by a statement not readily refutable for lack of ample

---

[56] *From Unity to Pluralism*, p. 194. Referring to their epistemological views on Thomistic realism, Maritain wrote: "Between M. Et. Gilson's position and ours there is no substantial difference," *The Degrees of Knowledge*, translated under the supervision of Gerald B. Phelan (New York: Scribner's, 1959), p. xvi. In turn, referring to the relation of reason and faith, Gilson advises: "On this general characteristic of Thomistic thought, see the basic work of Jacques Maritain *Distinguer pour unir, ou les dégres du savoir*." *The Christian Philosophy of St. Thomas Aquinas*, p. 443, note 49.

[57] Cf. *From Unity to Pluralism*, p. 194.

[58] Father McCool attributes to Gilson the verdict that "none of the contemporary forms of 'Thomism' was genuine Thomism." *Ibid*., p. 195.

[59] *Ibid*., p. 197.

statistics. If the statement is true, and granting that to some extent it may possibly be true, it is bound to be a disappointment to Gilson's numerous admirers to be faced with having to conceive of his tireless and devoted efforts to promote a living Thomism bearing such unintended and opposite results. Even so, should the Thomistic movement actually have reached a point of terminus (certainly it has slowed down), it would still be necessary to acknowledge that the end of the movement does not itself mean the end of Thomism.

In conclusion, I think we must return to the beginning. As Vernon Bourke pointed out,[60] there are two basic approaches to the thought of St. Thomas Aquinas: the so-called historical-textual approach and the approach of re-thinking the doctrine of St. Thomas in relation to contemporary problems and thought-patterns. A living Thomism requires both approaches. Gilson and Maritain were not alien to either of these approaches, although Gilson is more representative of the first approach, Maritain of the second. Both were faithful to the primary principles of St. Thomas in their interpretations of Thomism, although what they said about Thomism, as what anyone else says about it, must always be measured against the writings themselves of St. Thomas. Each of these Thomist scholars endeavored according to his own approach, his own manner of presentation, his own interest in contemporary problems, his own style of writing to make the teaching of St. Thomas vital and relevant in this twentieth century. Those who became Thomists under their tutelage will no doubt credit them with considerable success.

The teaching of St. Thomas which they so ably expounded is an intellectual heritage with ample resources to enrich humanity for ages to come. It may suffer temporary eclipse, as at intervals does the physical sun. Still, its light appears too brilliant and its wisdom

---

60 See note 2 above.

too indispensable ever to be dimmed to extinction. In the succession of future Thomists, certainly all will not be as gifted as Gilson and Maritain, yet somehow or other, Thomism, it seems, in virtue of its appeal to the sapiential instinct of the human mind, is destined to stay alive, at least in the halls of learning, if not equally so in the culture of the day.

NEO-THOMISM AND CHRISTIAN PHILOSOPHY

Joseph Owens, C.Ss.R.

I

Probably no writer today is more conversant with the historical details of Neo-Thomism, at least in Anglophone America, than is Fr. Gerald A. McCool. Looked at from the philosophical viewpoint, however, one of his conclusions is at first sight surprising. This one conclusion is worded: "On the basis of St. Thomas' own fundamental principles, the Catholic theologian now realized that the dream of the Neo-Thomistic movement could never be realized."[1] It seems like a provocative utterance. It would mean that Aquinas' own basic principles hold up a red light to the ongoing development of Thomistic thought in the present-day intellectual world. Further work in that direction would thereby be not only discouraged but positively blocked.

Of course a philosopher may always be grateful for, and gladly profit by, relevant information on what theologians are thinking about professedly philosophical work. Much serious philosophical effort is now being devoted to the progress of Christian philosophy in today's world.[2] In this effort the role that can be played by the philosophical thought of Aquinas, as envisaged in Leo XIII's encyclical on Christian philosophy, is given a prominent place.[3] To be told, then, that today's Catholic

---

[1] Gerald A. McCool, S.J., *From Unity to Pluralism: The Internal Evolution of Thomism* (New York: Fordham University Press, 1989), p. 228.

[2] See articles in the issue "Christian Philosophy," *The Monist*, 75:3 (July, 1992).

[3]*Acta sanctae sedis*, 12 (1879), 97-115. English translation of the encyclical may be found in *The Papal Encyclicals, 1878-1903*, ed. Claudia Carlen (Consortium Books: McGrath Publishing Company, 1981), pp. 17-26. The numbering used in that edition

theologian regards the aim of the Neo-Thomistic movement as something intrinsically unrealizable, and as contrary to the fundamental principles of Aquinas himself, comes with the ring of an oxymoron. Yet perhaps it may serve a useful purpose by suggesting a new and closer scrutiny of the meaning and worth of the continued effort to give Aquinas' thought its due place in the western philosophical enterprise.

The first question in this context is obviously that of the tribunal before which the case is to be heard. Is it a court of philosophy, or a court of theology? If the topic at issue is the Christian philosophy of Aquinas, the tribunal cannot be other than strictly philosophical. An ultimate decision in Christian philosophy, *qua* philosophy, has to be given by philosophy itself. This tenet was pressed home with force and precision in the well-known debate on Christian philosophy that had originated with Bréhier and Gilson. *Qua* philosophy, Christian philosophy is responsible solely to the court of unaided human reason. This norm was accepted unhesitatingly by both sides in the long debate.[4] The theologian's assessment of it *qua* philosophy, then, can have the status only of an outside and non-decisive view. For those concerned with assuring the philosophical thought of Aquinas its important place in the ongoing course of western philosophy, the verdict of a Catholic theologian is something of which notice may be taken. But it can have no probative force.

While a Thomistic philosopher has to read the works of Aquinas as theology, the further development of their philosophical aspects has to take place in each reader according the reader's own individual tendencies.

---

for the sections of the encyclical will be followed in the present article.

4 See Etienne Gilson, "La notion de philosophie chrétienne," *Bulletin de la Société française de Philosophie*, 31 (1931), 39; cf. pp. 42 and 49. For Emile Bréhier, *ibid.*, pp. 49 and 52.

The development will proceed in line with the philosophic habituation of the new thinker, without being dictated, or "managed," by theology. The development constitutes genuine philosophical progress, but the responsibility for it rests on the shoulders of the one who is engaged in it. As Anton Pegis stressed so forcefully, the progress comes about not by fragmentation of St. Thomas, but "by risking our own intellectual lives in the world of today."[5] Attempted management by theology has proven fatal to its activity as philosophy. According to the medieval simile, philosophy is absorbed into the wine of theology. The colloidal solution is wine, but in it the water retains its own substantial nature and can be evaporated without chemical change. While in the solution, however, it is not adapted to the ordinary uses of water. Considerations that are read in Aquinas as theology, then, can be extracted as philosophy by each individual thinker. But as extracted in that way and developed on the purely philosophical level, the thought has to stand on its own feet and avoid coloring or managing by theological interests, despite the inspiration and incentive given by Christian concerns.

II

With the way cleared for an assessment of Thomistic philosophy on the basis of strictly philosophical norms, one may ask if the facts

---

[5] "Let us admit that a theologically managed philosophy--philosophy used and shaped by the theologian in his world and for his purposes--is not philosophy." Anton C. Pegis, *St. Thomas and Philosophy* (Milwaukee: Marquette University Press, 1964), pp. 39-40. That would not be philosophy, but "a dead piece of theology," p. 41. Pegis concluded: "I hope that we shall undertake to build our own Christian philosophy, not by detaching fragments from his theology, but by risking our own intellectual lives in the world of today" (p. 88).

substantiate the conclusion expressed by McCool in the name of the contemporary Catholic theologian. On Aquinas' own fundamental principles is the "dream of the Neo-Thomistic movement" something that could never be realized?

There is much in the wording of this statement that calls for elucidation and precision. First, what exactly is meant by "the Neo-Thomistic movement," and what is its "dream"? The "dream" quite obviously means the purpose or result envisaged by the Neo-Thomists. As with the encyclical *Aeterni Patris* itself, the purpose was the restoration of Christian philosophy as the appropriate means for counteracting baneful effects of *modern* philosophies, namely the philosophies of the Enlightenment. In today's postmodern atmosphere, considerable effort is required to sense the impact of this approach. Philosophies emulating the success of the burgeoning physical and life sciences, with their demand for the same mathematical exactitude in all who pursued them, had been striving to control human destiny. From that angle they were tending to serve as substitutes for religion. But the nefarious results in fascist and communist totalitarianisms had not yet made themselves visible. The facts were not yet there to have their own convincing weight thrown into the discussion. The destructive philosophies had to be met with philosophy itself, and not with accomplished facts or revealed truths. It was a case of diamond required to cut diamond. The current movements had to be met on their own level, that of philosophy.

On this attitude, *Aeterni Patris* is surely clear enough. It looks back with approval upon Origen's use of tenets culled from pagan writers, as though snatching weapons from the enemy.[6] It compares the process with

---

[6] *Aeterni Patris*, no. 4 (p. 18b). Cf. no. 7, pp. 19-20.

the despoilment of the Egyptians.[7] It is looking at sources outside formal revelation. It continually uses the term "philosophy." There can hardly be any doubt that the term "philosophy" was being understood in the accepted division of the academic disciplines. The specific kind of philosophy recommended was the type that had been used by the medieval and Renaissance Scholastics.[8] Clearly, what was meant in this regard by "philosophy" was the discipline current in academic circles at the time under that title. There need be no hesitation in recognizing that the discipline whose restoration Leo was advocating belonged entirely to the area designated as "philosophy" in today's academic world.

Though the designation "Christian philosophy" was not used in *Aeterni Patris* itself, it was the wording in which the content of the encyclical was described a year later by Leo himself.[9] The encyclical focused its attention on this philosophy as it is found in Aquinas. But the philosophy was recognized widely throughout patristic and medieval writers. Augustine, Bonaventure, and Albert were highlighted.[10] It was a way of thinking that was viewed as permeating the Christian tradition. But Aquinas was commended in it for "his solidity and

---

[7] ". . . that by a change of use the things might be dedicated to the service of the true God." *Aeterni Patris*, no. 4 (p. 18b).

[8] *Aeterni Patris*, nos. 16-20 (pp. 22b-25a).

[9] The designation "On Christian Philosophy" was given the encyclical by Leo XIII himself on its first anniversary: ". . . ab Encyclicis Litteris Nostris *De philosophia christiana ad mentem s. Thomae Aquinatis Doctoris Angelici in scholis catholicis instauranda*, quas superiori anno hoc ipso die publicavimus." "*Cum hoc sit,*" *Acta sanctae sedis*, 13 (1980), 56.

[10] *Aeterni Patris*, nos. 6 (p. 19b) and 13 (p. 22a), for Augustine; no. 30 (p. 26a) for Albertus Magnus; no. 14 (p. 22b) for Bonaventure. Anselm is mentioned in no. 13 (p. 22a).

excellence over others."[11] Even within Thomistic circles varieties of interpretation were noted, along with "the established agreement of learned men."[12] The encyclical itself did not elaborate on the nature of these differences and the type of agreement. For us today this is a question of pluralism, a question to be settled on the genuinely philosophical level. It does not come under magisterial concern, and did not call for treatment by the encyclical. No, it is entirely a question of pluralism in Christian philosophy. Its solution has to be worked out by a study of philosophy as a human science, as science whose starting points differ from individual to individual and thereby allow for an indefinite plurality of philosophies. This holds for Christian philosophy as well as for any other kind. The faith may be the same for Christians, but the individual mental formation that each brings to the task is different in each. The philosophical study of pluralism is not to be "managed" by a desire for radical unity after the model of the unity sought by the Catholic theologian for Christian faith. Philosophical pluralism was not a problem of *Aeterni Patris.*

These burdensome considerations have been necessary for the approach to the present issue. McCool (p. 1) regards *Aeterni Patris* as the "*magna charta* of the Neo-Thomist revival." In its definite and towering statements, then, should the "dream of the Neo-Thomist movement" be probed, rather than in the extensive ranks

---

[11] *Aeterni Patris*, no. 31 (p. 26b). The Latin is "eiusque prae ceteris soliditatem atque excellentiam"--*Acta sanctae sedis*, 12 (1879), 114. This guarantees the "solidity" of the Thomistic doctrine, and against the Aristotelian background is open to interpretation in the *pros hen* sense. But no interpretation is given by the encyclical itself. That is a philosophic, not a magisterial, problem. Christian philosophy is regarded as one by the encyclical. But how it is one, does not come under the purview of *Aeterni Patris*. The instances mentioned are numerous and varied.

[12] *Aeterni Patris*, no. 31 (p. 26b).

of the individual predecessors or followers in each of which the interpretation may be expected to differ in accordance with her or his philosophical habituation. McCool, however, seems to take for granted that the "dream" or the "hope" of the Neo-Thomist movement was that of a monochrome "system" of philosophy that alone was right while all other "systems" were either wrong or merely approximations to the one correct "system." One could count numbers of Neo-Thomistic writers, even among the best, who held that view. If that is regarded as the "dream" or "hope" of the movement, the goal is obviously unrealizable. But decades ago this conception has been shown to be wrong.[13] It is nowhere to be found in *Aeterni Patris*. The encyclical aimed to restore Christian philosophy, as understood in the vast and variegated background of the whole of Christian tradition. It drew upon the wisdom of the centuries. It does not speak of Christian philosophy as a "system." In Aquinas, proposed as the model for Christian philosophy, there is quite obviously no "system" of philosophy at all. Whether the envisaged Christian philosophy is pluralistic or monochrome, is not a concern of the encyclical itself. That question is properly and rightly left for philosophical consideration. Probing of it belongs to a genuinely philosophical tribunal. In a word, the goal striven for in *Aeterni Patris* is Christian philosophy as pursued down the ages.

---

[13] The problem was faced at the annual meeting of the American Catholic Philosophical Association in 1966. In the presidential address that year I emphasized that no "system" was to be found in the texts. See "Scholasticism--Then and Now," *Proceedings of the American Catholic Philosophical Association*, 40 (1966), 7, n. 22. On the designation *philosophia perennis*, see my survey in "The Notion of Catholic Philosophy," *The Antigonish Review*, 1 (1970), 131-132, n. 8. The term "system" is not used in the Latin text of *Aeterni Patris*, though it occurs in English translations.

[14] See McCool, p. 225.

Whether that philosophy is pluralistic or monotone is not its concern.

If the *magna charta* of the Neo-Thomistic movement is seen in *Aeterni Patris*, then, no ground is offered for placing the "dream" or "hope" in a uniform "system" of philosophy. That one does, will rank only as an assumption. The facts do not substantiate the view. But it is also a fact that many Neo-Thomists, even among the highest ranking, have held it. This suggests a closer look at the ranks of those who may in one way or another be designated as Neo-Thomists.

In the widest sense, anyone in the nineteenth or twentieth century who worked for the restoration of Thomistic thinking in today's world may be called a Neo-Thomist, much as some may dislike the designation. But among these the pluralism is vast. The type of Neo-Thomism highlighted by McCool is a kind that proceeded from a tradition of able thinkers in the early decades of the present century. It was given broad Kantian grounding by Joseph Maréchal, and was "osmosed"[14] in the work of Bernard Lonergan. It made human insight and consciousness the immediately and directly known basis of philosophical inquiry. In this regard some may hardly hesitate to see in it the culmination of the whole Neoscholastic trend of thinking. From Thales on, even in Plato, Plotinus and Augustine, real things external to the knower had been the immediate and direct object of human cognition. Upon them, all further human knowledge was based. But with Descartes a drastic revolution took place. Philosophy had to be based upon human ideas only, with the notion of "idea" extended to sensations in Hume and Condillac. With a few notable exceptions this stance prevailed throughout the Neo-Thomistic revival, as may be seen by a glance at the history of its epistemology.[15]

---

[15] See Georges Van Riet, *Thomistic Epistemology*, 2. vols. (St. Louis: B. Herder Book Co., 1963-1965); E. Gilson,

Its proponents were aware of being locked up within their own cognition, and sought despairingly for a "bridge" to enable them to reach the outside world. But no satisfactory bridge was ever found. If this Cartesian aspect of Neo-Thomism is made the norm for assessing the whole movement, then patently McCool's interpretation of it is correct.

But the facts show no such logical sequence. The remote background of the movement was much deeper than Descartes. In fact, a concerted effort is required to show that Neoscholasticism was basically a new brand of Cartesianism. *Prima facie* its main philosophic source appeared to be Aristotle, whether or not its proponents had ever read Aristotle in the original. It was to Aristotle's works that they continually referred. It was Aristotle's terminology that they predominantly used, and Aristotle's broad divisions of philosophy that they followed. But as regards the all-important question of philosophical starting points, they were not basically in his world. A philosopher may locate his starting points in external things, in human thought, or in human language. In the ancient and medieval worlds, the starting points were external things immediately and directly known. In modern and Enlightenment philosophy they were the human mind and its cognitional content. In both Continental and analytic philosophy, they are found in human language. Viewed from this angle, the Aristotelian starting points were external sensible things. The knower and the cognitive activity were grasped only concomitantly and indirectly.[16] In their Aristotelian sources, accordingly, the Neo-Thomists were facing in fact a philosophy in which all human cognition originated in external sensible things,

---

Thomistic Realism and the Critique of Knowledge, trans. Mark A. Wauck (San Francisco: Ignatius Press, 1986).

[16] See Aristotle, *Metaphysics*, 12.9.1074b35-36. Cf. Nicolai Hartmann, *New Ways of Ontology*, trans. Reinhard C. Kuhn (Chicago: H. Regnery, 1953), pp. 16-20.

rather than in the cognitive agent and activity that could be grasped only *in obliquo*. Ideas, or sensations, or consciousness could not serve as the original basis for philosophical procedure.

In mixing the Aristotelian with the Cartesian procedures there was of course inconsistency on the part of the Neoscholastics. But that same inconsistency was found everywhere in the philosophies that developed in the wake of Descartes. Neither Descartes, nor Locke, nor Hume had any real doubt about the existence of the external world, nor has anybody else outside a mental institution. They were all certain of its existence, but they had to look elsewhere, outside their own philosophical starting points, for the ground of that certainty. Even analytic philosophers are aware of a reality against which they check the correctness of their analysis, and hermeneuts of a reality they are trying to reach as best they can in their linguistic interpretations. In this inconsistency, then, the Neo-Thomists were in perfectly good company. Like everyone else they lived in the real world and could not shake that knowledge from their philosophizing. As Gilson noted, the external world like Caesar's wife has to be above suspicion.[17] But this should not have dispensed them from explaining on Aristotelian philosophic principles how the external thing succeeded in getting into human cognition. With Aristotle the thing known and the cognitive agent were one and the same in the actuality of the cognition, with the external sensible thing known directly and the cognitive agent only concomitantly. But the Neo-Thomists failed to exploit this rich epistemological source. Nor did they note that direct self-knowledge, according to their Aristotelian sources, would be restricted to one's self as in the separate substances, with no possibility of knowing anything else. There would be no question of a "bridge" to an outside world.

---

[17] E. Gilson, *The Unity of Philosophical Experience* (New York: Charles Scribner's Sons, 1937), p. 184.

The facts, then, show that the Neo-Thomistic movement was open to development in many directions. It was not doomed to follow the Cartesian tendency that would logically bring it to a consummation in Transcendental Thomism or any other kind of Thomism. It remained wide open, on account of its use of reciprocally inconsistent sources. The sources recommended by *Aeterni Patris* were historically antecedent to the Cartesian deviation. They were seen in Aquinas himself, read in company with the Christian wisdom that had preceded him through the centuries. In Aquinas, all philosophical knowledge arose from external sensible things as existent in themselves, rather than from ideas or sensations of them. The really existent things came, through human cognitive activity, to exist cognitionally in one's mind. Things existing in themselves were directly grasped, and upon them all further philosophical reasoning was based. But the existential factor was stressed, in contrast to the Aristotelian presentation. This left the interpretation of the Thomistic texts open to bewildering confusion that has persisted to the present day, and in which the defenders of Aquinas came to speak the language of Aquinas' adversaries.[18]

That complicated history of the Neo-Thomistic movement needs to be kept in mind when assessing its agreement or dissonance with "St. Thomas' own fundamental principles." Actually, did the hope of this movement have "no real support in the philosophy of St. Thomas himself?"[19] If the Neo-Thomistic movement is held to be basically the trend that culminates in Lonergan, then of course it is diametrically opposed to Aquinas' own fundamental principles. The basic division in the western philosophical enterprise is

---

[18] Rolf Schönberger, *Die Transformation des klassischen Seinsverständnisses* (Berlin: Walter de Gruyter, 1986), p. 386; cf. p. 13.

[19] McCool, pp. 228-229.

between philosophies that start from external things and those that start from the internal activities of either thought or language. In that utterly fundamental division the starting points of Aquinas himself are external sensible existents, while the starting points of the Neo-Thomistic movement as signalized by McCool lie within human insight and consciousness. Viewed from the Cartesian dividing line the opposition between the two could not be more stark. If Neo-Thomism is regarded as restricted to or even dominated by the structure erected upon the notion of insight into insight, then wholehearted agreement with McCool's conclusion is logically the result. One may unhesitatingly grant that this type of thinking "had no real support in the philosophy of St. Thomas himself," and was in this way radically anti-Thomistic. But the strident oxymoron in the proposition that the Neo-Thomistic movement is anti-Thomistic is no longer present. The view is not even provocative. Rather, it has the status of a normal and unpolemical observation.

III

But is there any solid ground for interpreting the Neo-Thomistic movement as a uniformly directed philosophy that of its nature tended towards the way of thinking reached in Lonergan? Certainly the great majority and the most widely read of the Neo-Thomistic writers did not show signs of that tendency. By and large, rather, they followed the Aristotelian propensity of looking to external sensible things for their starting points, in spite of the fashionable Cartesian influences that inspired distrust in regard to knowledge of sensible things. One may easily regard this attitude as inconsistent, but it was the common attitude of the day. It resulted in the pluralism that was present in the Neo-Thomistic movement from the start. Each individual Neo-Thomistic writer, at least insofar as she or he was not merely chopping off fragments from the rich

Thomistic corpus, was interpreting Aquinas in line with a personal background. These writers may be grouped according to "family resemblances," quite as DeWulf had claimed in regard to their medieval predecessors.[20] But they are far from monochrome in character. In the *magna charta* of the movement, namely the encyclical *Aeterni Patris*, there is no mention of a *philosophia perennis* that would be univocally the same for all, nor was there even an allusion to this conception of philosophy as proposed by Steuco and Leibniz. Many of the Neo-Thomistic writers did in fact envisage their work along the lines of a *philosophia perennis*, but others objected strongly to the notion. This is but another indication of their pluralism. Their *magna charta* neither required nor forbade that uniformity. Rather, *Aeterni Patris* looked to the entire panorama of patristic and medieval writers for the Christian philosophy it aimed to restore. Pluralism it allowed, and pluralism was everywhere present.

Again, one may stress that *Aeterni Patris* was not interested in the pluralistic or non-pluralistic nature of Christian philosophy. It was envisaging Christian philosophy as something that had been patently active in patristic and medieval writers. It was ardently advocating the restoration of that philosophy. Even after surveying the various religious orders and European universities in regard to the philosophy taught in them, the encyclical was able to conclude: "Thomas reigned supreme."[21] That was its attitude towards the

---

[20] See Maurice de Wulf, *Scholasticism Old and New*, trans. P. Coffey (London, 1910), p. 46, for the notion of "family resemblances." In his *History of Medieval Philosophy*, 3rd ed., trans. P. Coffey (London, 1909), p. 109, De Wulf recognizes the individuality of each of the Scholastics, comparing this to the various members of a single family.

[21] *Aeterni Patris*, no. 20 (p. 23b). Against the Aristotelian background, this lends itself easily to a *pros hen* interpretation. But the encyclical is not interested in that point. Its explicit

philosophy it meant to promote. There should be little wonder, then, at finding even more extensive pluralism later on among the Neo-Thomistic writers, in accord with the wider background in the recent philosophic world.

IV

Wide possibilities of interpretation are, in consequence, open to today's student of Neo-Thomism. On what basis is the choice to be made? The Catholic theologian, for whom McCool is speaking, traditionally tends to proceed from the viewpoint that philosophy is the handmaid of theology. In that way the selection will be made according to the help that philosophy can offer to today's Catholic theology. In large part this theology aims at modernizing its teaching in a way that will conform to present-day conditions. It tends, consequently, to regard as outmoded anything that is medieval. This, obviously, includes the way Aquinas thought. While Aquinas may have been abreast of his own times in using Aristotle to elucidate Christian thought, he is not properly adapted to the objectives of modern thinking. Kantianism, Hegelianism, phenomenology, existentialism, and, with some liberation theologians, even Marxism are much better suited for the purpose. The Catholic theologian today, then, should choose the type of philosophy that best assists this modernization process and tendencies. That is the proper function of the handmaid in today's circumstances. This helps to understand how a type of Transcendental Thomism could be chosen as the dominant or characteristic aspect of the Neo-Thomistic movement.

---

doctrine, however, is that in the Christian philosophy to which it was referring Aquinas was at the peak, in whatever way you care to explain that supremacy.

But the function of handmaid is only one aspect of a philosophy. As philosophy it has its own fundamental nature antecedently to any help it can give the theologian. Philosophy, in fact, was in existence long before the advent of Christianity. Within Christian philosophy it does not lose its nature. It can be chosen for its own intrinsic merits. In fact, for the purpose envisaged by *Aeterni Patris*, that of diamond cutting diamond, its strength has to be exercised on the plane of natural reason. If its function as a handmaid to Catholic theology is used as a prop, the purpose is undermined. Theological management of this type would diminish its credibility among non-Christian philosophers. To attain its proper purpose, it has to be answerable solely to the tribunal of human reason. Ability to help towards the modernization of Catholic theology need not be used as a criterion for the choice of a handmaid.

Moreover, is that goal of modernization even an acceptable one today? We live in the postmodern era. The modern era was the age of the Enlightenment. It began with Descartes' insistence on human ideas as the ultimate basis for philosophical thinking, and with Bacon's projection of organized work among groups of scientists, with both factors developed on the mathematical model. For this type of work the same multiplication table and the same exact measurements had to be used by all. It resulted in the magnificent accomplishments of the modern world. But it began to cause dissatisfaction when used in the social sciences, where the factor of human freedom was paramount. Sociobiology did not work. The result has been conceptions of philosophy in which each person's philosophical thinking is as distinctive as her or his fingerprints or DNA, and in which the grounds are offered for today's philosophical pluralism. This is postmodernism.[22] It renders the goal of

[22] On the range and nature of postmodernism, see Jean François Lyotard, *The Postmodern Condition*, trans. Geoff

"modernization," in the sense of a theology conformed to the Cartesian and Baconian ideals of science, something quite out of date.

In postmodern pluralism, accordingly, one is free to choose the starting points upon which the relevant philosophical thinking is based. The fact that these starting points fit in with those held by writers in former ages need not be an obstacle. Chronolatry can hardly be allowed to enter as the deciding factor. Family

---

Bennington and Brian Massumi (Minneapolis: University of Minnesota Press, 1984). On *types* in postmodernism, see *Postmodern Genres*, ed. Marjorie Perloff (Norman: University of Oklahoma Press, 1988), pp. 3-7. For a short synopsis of the current concept of postmodernism in philosophy, see Joanne B. Waugh, "Heraclitus: The Postmodern Presocratic?" *The Monist*, 74 (1991), 609-612. A person's habituation in Christian culture is what makes Christian philosophy a distinct philosophic species, and sacred theology has played a notable part in the shaping of that culture. In this way sacred theology exercises a guiding role without entering into the principles of Christian philosophy itself. It merely leads up to the starting points in things, thought or language, as the Aristotelian dialectic does in regard to philosophy. This is quite understandable in the postmodern setting. Accordingly in works of the early sixties Gilson stressed the influence of sacred theology upon Christian philosophy. But this in no way changed the stand expressed by him in the thirties, that *qua* philosophy Christian philosophy is responsible solely to the court of human reason (see *supra*, n. 4). In his *Introduction à la philosophie chrétienne* (Paris: Vrin, 1960), p. 18, a philosophical conclusion is regarded as true in virtue of its rational ground, as contrasted with the ground in faith. On the formal distinction from sacred theology, see *ibid.*, pp. 112-123. In *The Philosopher and Theology*, trans. Cécile Gilson (New York: Random House, 1962), p. 179, Christian philosophy is still "truly rational," though "quickened by a genuinely Christian spirit." That fits into the postmodern framework, where each philosophy is specified in accord with the individual thinker's cultural formation. In consequence the relation between philosophy and sacred theology is not at all on a par with the relations of the philosophical sciences among themselves (p. 99).

resemblances to Aristotle in antiquity and to Aquinas in the middle ages are consequently no deterrent. From the philosophical viewpoint the issue here is to *understand* clearly what McCool is doing and what his motivation is. In a context where the motivation is philosophical, the task is to place the Neo-Thomistic movement in the broad panorama of western philosophy as a whole. As noted above, this enterprise has located the starting points for philosophical thinking in three readily distinguished areas. These are external things, human thought, and human language. From Thales on, even in Plato, Plotinus and Augustine, the starting points were located in one way or another in things themselves, other than the knower. From Descartes' time they have been located in human thoughts or sensations. With the Continental and analytic thinkers they came to be located, again in one way or another, in human language. Where the objective is to locate the Neo-Thomistic movement correctly in this wide sweep of the western philosophical enterprise, instead of assessing its ability to function as a handmaid to theology, the grounds for the decisive judgment have to lie in the way these three different types of starting points are seen to function in western philosophical thought.

What prospects do these three radically different ways of basing philosophy offer respectively for the interpretation of the Neo-Thomistic movement? The Enlightenment way of using human ideas or sensations as the starting points, occasioned among Neo-Thomists the futile search for a "bridge" to reality. None was ever found. As with an Aristotelian separate substance, there was no way in which the cognition could ever get outside itself and reach anything else. Yet the Neo-Thomistic thinkers claimed all along to be dealing with real things. They knew those things through their life in the real world, and they had to smuggle covertly that knowledge into their philosophical reasoning. They could not bring it in overtly, because against the Cartesian background this would mean admitting direct knowledge, in an immediate sense, of things in the outside world. That was not fashionable or even respectable at a time when it was branded as naïve realism. The absurd "things in themselves" could not be recognized overtly as the correct basis for philosophical thinking. Yet the Neo-Thomistic movement demanded everywhere that philosophy should bear upon real things. The Cartesian starting points, in consequence, proved incapable of leading to the objectives of Neo-Thomism.

The locating of the starting points in human language likewise did not have any success. There was still the problem of getting from language to reality. Yet even where analytic and hermeneutic techniques were used in the Neo-Thomistic movement, it still wished to be dealing with the real world and covertly assumed the existence of a world outside the speaker.

The case, then, falls back upon the Aristotelian background that the Neo-Thomists could not help acknowledging. In it knower and thing known were one and the same in the actuality of cognition. In that actuality they were thoroughly identical with each other. To know the one was to know the other. But there was an order in that actuality. The cognition was of something other than itself, and only concomitantly of its own self. There was no possibility and no need for a bridge to reality. The knower was already there, in the actuality of the cognition. If there were question of a bridge, it would be from the real external thing to the internal cognition, insofar as all form and shape came to the cognition from the external thing. Of itself the human intellect was pure potentiality. It received everything in the line of form or actuality from the external sensible things, by means of their efficient causality upon the human cognitive agent. This thoroughgoing potentiality of the human intellect was strenuously denied by writers in the transcendental tradition signalized by McCool, but there is no doubt that it was firmly expressed in Aristotle. It was what

Aquinas himself found in Aristotle, and what he developed incisively in terms of his distinction between thing and being. What is identically the same thing can have different ways of being. It has being in reality outside the mind, and cognitional being inside the mind. The sensible thing's form was received by the mind, and, as cause of being, this form made the cognitive agent be the external thing in the new cognitional being.

When one looks today for the dominant and characteristic strain in the onflowing Neo-Thomistic movement, then, the Aristotelian rather than the Cartesian or Kantian influence seems to prevail. Undoubtedly there is inconsistency here, but in fact the vast majority of Neo-Thomistic spokesmen do not at all sound like Descartes or Kant. They have taken too much, directly or indirectly, from Aristotle. In this regard one need not be frightened away by accusations of naïve realism. The doctrine as found in Aristotle was profound and penetrating. It was not at all naïve in its elaborate and deeply explorative inquiry into human cognition. Nor was it at all a realism, if realism is understood as the reasoning to the existence of the external world on the basis of human ideas. Yet that is how "realism" has been taken in the philosophies that expressly characterize themselves as "realisms." If the term were taken to mean the belief that we live in a real external world, then everybody would be a realist. In philosophical use it means that the existence of the external world can be proved philosophically, and in the Aristotelian context that is neither a requirement nor a possibility.

To sum up, then, the key to the correct interpretation of the Neo-Thomistic movement is to be found in the philosophy of Aristotle, when the question is placed in the general panorama of western philosophy and is not left to the function of philosophy as the handmaid of theology. It is not a question of "the mind's orientation to God's infinite intelligibility" as the guarantee for "the first principles of metaphysics[23] much as that may mean in a Maréchalian context. Rather, on the philosophical level it is a question of guaranteeing the basic principles of metaphysics in priority to philosophical knowledge of God. In that order only can Christian philosophy counteract atheism on the philosophical level, as envisaged in *Aeterni Patris*. In the Aristotelian background the being that manifested itself in sensible things guaranteed those basic principles, insofar as it showed that a thing could not be and not be at the same time in the same respects. In the interpretation of the Thomistic movement, accordingly, the Aristotelian background should be the paramount factor on the philosophical level for assessing the ability of that movement to play the role envisaged for Aquinas in the *magna charta* of Christian philosophy. Human insight or consciousness cannot function as basic object, for it has the status of a predicamental accident, and not of a thing in itself.

V

How, then, is McCool's assertion to be evaluated when it is understood against that Aristotelian background? How can one possibly hold that "St. Thomas' own fundamental principles" prevent "the dream of the Neo-Thomistic movement" from ever being realized?

First, this statement is made in the name of "the Catholic theologian." It is not attributed to the Catholic philosopher, and accordingly is not presented as framed in Christian philosophy. But Christian philosophy was the theme of the encyclical that is recognized as its *magna charta*. Christian philosophy has its own life to lead, and *qua* philosophy is accountable solely to the tribunal of human reason. Attempted management by theology is not good for its health. But Catholic theology has the

---

[23] See McCool, p. 228.

full right to choose as its handmaid any type of philosophy that it wishes. If it chooses to select Transcendental Thomism, that is its own affair. Suitable domestic help may be hard to come by in these days, and if it finds that type of Thomism congenial it can take it along with the consequences. One of those consequences will be that "St. Thomas' own fundamental principles" come into open conflict with the Neo-Thomistic movement, insofar as those principles are Aristotelian in cast rather than Cartesian or Kantian. Taken in the latter way, they would hinder the Neo-Thomistic hope from ever being realized.

Here Christian philosophy can only give advice. It cannot dictate. It can only show on its own level how a philosophy based on thought or language is radically different from a philosophy based on external sensible things immediately and directly known. It has no mandate to determine which type Catholic theology should select as the handmaid. But in *Aeterni Patris* it does have a mission to uphold and develop the philosophy handed down in the works of Aquinas. That is the dream and hope beckoning it on to future work.

The proposed task, then, is far from being a "rearguard action" that futilely strives to delay as best it can the surging modernization of Catholic thought. Rather, it regards that modernization as already outdated and as belonging to the Enlightenment mentality, a mind-set now superseded by the postmodern views that see philosophic starting points chosen in accord with one's cultural upbringing and personal habituation. If that habituation is Christian, the philosophy prompted by it will be Christian philosophy. The "Christian" factor will not furnish the starting points, but will incline the individual thinker towards them, be they in external things, in human thought, or in language, in order to generate philosophy helpful for Christian life. That was the kind of philosophy desired by *Aeterni Patris*, and its most suitable instance was located in the writings of Thomas Aquinas. There the philosophical starting points were, as in Aristotle, really existent external things, but now regarded as actuated by an existence that was not part of their essence and had to be received from outside.

It is this type of Christian philosophy that needs to be promoted strongly for the future. Work towards it is an advance, very positive, and not at all a "rearguard action." It opens out on a wide array of fields. The basic epistemological question, namely how things external to cognition can be thoroughly identical with the knower in the actuality of the cognition, still calls for careful study. The tenet that the knower becomes and is the thing known, instead of just having it, is still found very difficult for students to accept. They see that it is undoubtedly present in the text of Aristotle, but they offer great sales resistance to accepting it for themselves. The role of existence in demonstrating the existence of God, and its role in individuating creatures, still call for much probing. The omnipotence of God, as based upon his nature as existence without any limitation by essence, and its role in the act of creation and of free action in spiritual creatures calls for extensive study. The same can be said for the indestructibility of the human soul, in face of the difficulties about its immortality that troubled Cajetan. Important philosophical questions about the nature of the whole moral order, and about natural law and the extent of respect for human life, loom up with urgency today for students of Aquinas. There is a tremendous amount of work lying ahead for the Neo-Thomistic movement. These are only a sampling of the instances.

Further, the instruments for Neo-Thomistic progress are at hand today as never before. Busa's *Index thomisticus*, invaluable for location of Thomistic passages, and the Leonine critical texts, now offer their help. Even the commentary on the *Metaphysics*, so crucial for understanding Aquinas' contact with his Aristotelian sources, should be available shortly. From these angles all looks good for the future of the Neo-Thomistic movement as pursued in accord with the guidelines of *Aeterni Patris*. Nor should today's low level of reception on the part of the reading public prove an insuperable obstacle. Aristotle has throughout the long history of his philosophy repeatedly experienced rise and fall in popularity, yet at the present time is inspiring eager young philosophy students with an ardor that bodes well for the future. Aquinas has continually had his ups and downs, with euphoria in the early fourteenth century at the time of his canonization, and later at the use made of him in the sixteenth century at the Council of Trent, and then through the Leonine encyclical in the nineteenth century. After each of these bursts of attention he receded to a much lower level of notice. There is no reason to think that this alternating history will not be continued.

In regard to the question presently under discussion, then, the task of Christian philosophy is clearcut. Its problem is to show how anyone could possibly hold the suicidal notion that the fundamental philosophic principles of Aquinas are destructive of the Neo-Thomistic movement itself. The reason why this conclusion could be held becomes obvious enough. If you think that Neo-Thomistic philosophy is ultimately based upon human insight or consciousness, you are already on the Cartesian side of the profound chasm that lies between this way of philosophizing and the way that is based upon externally existent sensible things. In the overview of the western philosophical enterprise the radical division is that clear. It is not at all difficult to see how McCool's conclusion follows logically from that conception of the Neo-Thomistic movement.

From the viewpoint of Neo-Thomism as a type of Christian philosophy, moreover, its task has never been the mediation of "the Church's Scriptural and Patristic heritage . . . through a single perennial system of theology."[24] In performing that task it has already

[24] See McCool, p. 228.

been absorbed into the wine of sacred theology. For that ancillary function the choice and direction of the handmaid understandably belongs to theology. That kind of work is theological in character. But strictly on the level of Christian philosophy, the Neo-Thomistic movement can go about its task without any reference to the mediating function. It stands on its own feet as a type of philosophy in the postmodern age. Viewed from the Aristotelian background that historically molded the philosophical thinking of Aquinas, it warrants the conclusion that Neo-Thomistic philosophy may look forward to a very worthwhile future as the dominant type of Christian philosophy envisaged in the *Aeterni Patris* of Leo XIII. The question whether or not it is to be used as the handmaid, however, pertains to the Catholic theologian. The handmaid still has her own life to lead, even when discharged by her erstwhile employer.

THOMISTIC PHILOSOPHY IS NOT PLURALISTIC

Vernon J. Bourke

Every few years a philosophical book appears that is important because it demands a response from concerned readers. Such is the case with Gerald A. McCool's *From Unity to Pluralism: the Internal Evolution of Thomism* (1989, 1992). My reaction to this book is mixed. In my view it is not Thomistic philosophy that has pluralized but rather that some Catholic philosophers have abandoned Thomas Aquinas and followed diverse ways.

There is no question that McCool's book offers a careful appraisal of the thinking of four leading Catholics who lived in the twentieth century. Pierre Rousselot, S.J., Joseph Maréchal, S.J., Jacques Maritain, and Etienne Gilson are depicted as key figures in the development of Roman Catholic thought following the publication of the famous encyclical *Aeterni Patris*. It is McCool's thesis that this succession of thinkers represents an "evolution" of the unitary doctrine of the encyclical into a pluralistic situation in the later decades of our century.

These four Francophone writers are called "thinkers" above in order to avoid saying whether they are primarily philosophers or theologians. The first two were priests and the second two were Catholic laymen. All four were remarkable scholars but it is not easy to decide on their major fields. As I see them, the first two were theologians and the two non-clerics were philosophers who dabbled a bit in theology. That is one reason why my reaction to McCool's presentation is mixed. I am not a theologian: by that I mean that I never studied sacred doctrine in a formal course. (Nor did Augustine of Hippo, yet he is sometimes called a theologian, though he thought the *theologi* were pagan intellectuals.) However, I have read and admired some patristic and medieval theologians and I have seen, but not always admired, the work of some recent theologians.

The problems raised by McCool's book are many. I propose to examine five of them. 1) Is there a standard meaning for the word philosophy? 2) Is there a standard meaning for theology? 3) What is the difference between philosophy and theology? 4) Is there room for a "Christian Philosophy"? 5) Is Thomistic philosophy pluralistic? My comments will be restricted to the role of Etienne Gilson in the so-called pluralization of Thomistic philosophy, for I do not feel competent to deal with the other three writers.

I. The Meaning of Philosophy

My old *Webster's Collegiate Dictionary* (1948) says that philosophy means: "Literally, the love of wisdom, in actual usage, the science which investigates that facts and principles of reality, and of human nature and conduct . . . ." Later it names its parts: "logic, ethics, aesthetics, metaphysics and theory of knowledge." In D. D. Runes' often lamented *Dictionary of Philosophy*,[1] J. K. Feibleman wrote that philosophy is: "Originally the rational explanation of anything; the general principles under which all facts could be explained. . ." He added: "technically, the science of sciences, the criticism and systematization or organization of all knowledge . . ." To my mind the Webster definition is better.

It is helpful to recall what the first recorded Greek philosophers thought they were doing. From Thales on, they criticized the myths of pagan religion as found in Hesiod, Homer and popular folklore, and they tried to find a wiser set of answers to questions such as: what is the prime stuff in the physical universe, what is more basic, becoming or being, and what is the nature of

---

[1] New York: Philosophical Library, 1942, p. 235.

human virtue? My first three years of teaching at Toronto were devoted to Greek philosophy. I have little doubt that these first philosophers were critical of popular religion and dissociated themselves from it. After almost a thousand years the last of the classic Greek thinkers (such as Plotinus and Porphyry) attempted a sort of reconciliation of philosophy and theology. Fusion may have led to confusion.

In our century, philosophers gather every five years under the direction of an international organization[2] to present papers, presumably philosophical, and face criticism. My experience of these quinquennial confrontations is that there is no one accepted meaning of the term "philosophy." It always amused me to overhear the conversation of an English analytic philosopher with a continental European phenomenologist. Apart from difficulties of diverse languages, there was an obvious lack of community of thinking. Marxist philosophers read papers hardly up to the level of high-school physics. One year (was it at Brussels or Venice?) a Swiss thinker sent in a paper entitled, "The Philosophy of Zero." At the FISP pre-congress meeting this nihilistic effort was discussed and rejected. The Swiss writer had his paper privately printed and circulated. The incident left the impression that there are some limits to admission to the domain of philosophy.

In 1992 there were 6,790 members of the American Philosophical Association.[3] Most are teachers--but of what? Textbooks are no longer in style in the classroom. Books of selected readings from the history of philosophy are often used as jumping-off points for the philosophy instructor's own opinions. As a result there are in this country more than six thousand

---

[2] FISP is the Fédération des Sociétés Philosophiques.

[3] See Hoekema, D. A., "Philosophers in the United States: Who Are We and What Are We Doing?" *Proceedings and Addresses of the American Philosophical Association*, 65, 7 (June, 1992), 41.

kinds of "philosophy" being taught. At times I have been required to evaluate the work of students transferring from one institution to another. It is almost impossible to judge whether the course taken by Joe College at one school, in ethics, philosophy of mind, and so on, is the equivalent of the same titled course at another institution. This is pluralism run rampant. When Americans started a fraternity dedicated to the notion that philosophy is the helmsman of life (*Phi Beta Kappa*), they all had much the same meaning for philosophy. That is no longer true.

Granting that there is no universally accepted definition of philosophy today, one may still say that it is the attempt to solve the ultimate problems that arise from ordinary human experience and from hard questions in the special sciences. These problems include: the nature of correct thinking or reasoning; the cause-effect relation including the question of a first cause of all reality; what it means to exist and to be some kind of thing; the difference between general understanding and the perception of particular objects; the meaning of goodness, unity, truth and beauty; the discovery of what it takes to live a good and happy personal life; the characteristics of the best possible human society.

Such a descriptive statement would, I think, be approved by a good many professional philosophers today. Some would wish to add or subtract from the enumeration of problems, but they would not claim that philosophy is just another empirical science, nor that it is the rational explanation of some religious faith. Philosophy is not just some meta-science studying the other sciences.

II. The Meaning of Theology

Of course the etymological signification of *theologia* is speaking or reasoning about God. On the other hand, the notion of "faith seeking understanding," which permeates patristic and medieval writing (whether

Jewish, Moslem, or Christian) is still a simple and respectable way of describing theology. The act of faith, believing, is the personal acceptance of some affirmation as true, because it is vouched for by a respected authority, even though it is not seen or assented to as clearly evident.

III. The Difference between Philosophy and Theology

Since this is written primarily for readers who call themselves Thomists, or at least have some interest in the thought of Thomas Aquinas, may I say immediately that we must clearly distinguish philosophy from theology. It is often said that the philosopher moves by reasoning from lower types of knowing to the understanding of higher truths, while the theologian starts with higher truths and uses them to explain what is lower. More informative is what Thomas says in the first article of his *Summa Theologiae*:

> Sciences are diversified according to the diverse notions of their knowable objects. For the astronomer and the physicist both prove the same conclusion--that the earth, for instance, is round: the astronomer by means of mathematics (i.e., abstracting from matter), but the physicist by means of matter itself. Hence there is no reason why those things which are treated by the philosophical sciences, so far as they can be known by the light of natural reason, may not also be treated by another science so far as they are known by the light of divine revelation. Hence the theology included in sacred doctrine differs in genus (*differt secundum genus*) from that theology which is part of philosophy.[4]

---

[4] *S.T.* I, 1, 1, ad 2m; as revised by A. C. Pegis in *Basic Writings of St. Thomas Aquinas*, (New York: Random House, 1944), I, p. 6.

This means, I think, that there are two ways of studying basic human problems and they differ generically, that is, in the broadest possible manner. The theology that is better called sacred doctrine (*sacra doctrina*) is quite different from the "natural theology" that is called metaphysics by Aristotle, which is, of course, philosophical. This is the same distinction that Thomas had made a decade earlier in his commentary on the *De Trinitate* of Boethius. There Thomas said:

> Accordingly there are two kinds of science concerning the divine. One follows our way of knowing which uses the principles of sensible things. The other follows the mode of divine realities themselves . . . Thus the truths we hold on faith are, as it were, our principles in this science, and the others become, as it were, conclusions. From this it is evident that this science is nobler than the divine science taught by the philosophers, proceeding as it does from more divine principles.[5]

It is well to remember that science (*scientia*) has, for Aquinas, a broader meaning than the term is given today. In Thomas' writings it means any kind of well-ordered knowledge, especially that using reasoning from initial premises (*principia*) to rational conclusions. So, *sacra doctrina* or *scientia divina*, as starting from faith in what God has revealed to the human knower, is clearly different from philosophy. Nor, to my knowledge, does St. Thomas insert any intermediary science between theology and philosophy. So, when we talk about "Thomism" and its possible evolution, it is advisable to make clear whether we are looking at theology or some kind of philosophy. This distinction is not always plainly made by McCool.

---

[5] *Quaest*. 2, 2, resp.; trans. Armand Maurer, St. Thomas Aquinas, *Faith, Reason and Theology*, Questions I-IV of his commentary on the *De Trinitate* of Boethius, (Toronto: Pontifical Institute of Mediaeval Studies, 1987), pp. 41-42.

If it is the case that philosophers today differ widely on the meaning of their craft, it seems to me that there are many ways of doing theology. What most theologians seem to be doing is explicating (unfolding) the contents of their beliefs, showing what are the implications of things known from revelation. Some use philosophy as an aid; others use psychology, history, social science, linguistics, personal meditation, and so on. In any case it seems clear to me, as it did to Aquinas, that theology is quite different from philosophy.

IV. The Problem of a Christian Philosophy

Both Jacques Maritain and Etienne Gilson lived in a period in Europe when Catholic scholars had a hard time gaining acceptance in faculties of philosophy at the great universities not under Catholic sponsorship. From the 1920's onward there were bitter disputes between French Catholic philosophers and their critics.[6] When Gilson first (1927) visited St. Michael's College in the University of Toronto, I was a third-year student in the honor philosophy class. This famous French scholar was already on the defensive in regard to Christian philosophy. Gilson told me, when I was working for my master's degree in 1929, to write my thesis on the Christian philosophy of St. Bernard! The conclusion of the paper I submitted was that there is no philosophy in Bernard's writings. Eventually Gilson published a book on the Mystic of Clairvaux. He kept Bernard in his *History of Christian Philosophy in the Middle Ages* (1955) but his earlier book dealt with the mystical theology of St. Bernard.[7]

---

[6] Cf. L. Shook, *Etienne Gilson*, (Toronto: Pontifical Institute of Mediaeval Studies, 1984) pp. 131-201.

[7] See *La Théologie mystique de saint Bernard*, (Paris: Vrin, 1934), trans. as *The Mystical Theology of St. Bernard*, by A. H. Downes (London-New York: Sheed & Ward, 1940).

I do not think Gilson ever located a writer in the history of philosophy that would pass muster as a Christian philosopher in the Gilsonian sense of the term. No doubt St. Thomas was his favorite thinker but Gilson always stressed the point that Aquinas was a theologian. In his classes and writings Gilson used Thomas' theological works almost exclusively: the commentaries on Aristotle that Aquinas took years to produce were rarely mentioned.[8]

Gilson never claimed to be a theologian. What he maintained was not that Christian philosophy depended on theology; rather, he always insisted that what made a philosophy "Christian" was its relation to faith. Certain germinal themes in Scripture (such as the Exodus 3:14 text where God says, "I Am who Am,"[9] and the opening lines of St. John's Gospel, where the starting-point [*principium*] of all things is named the Word of God bringing life and light to all[10]) were for Gilson the seeds of philosophical wisdom.

As early as 1935 Gilson published an important statement of his views on the relation of philosophy to the other academic disciplines. This was his essay on "The Distinctiveness of the Philosophic Order."[11] After suggesting that each kind of science, "by reason of its distinctiveness [requires] an appropriate mode of

---

[8] Cf. McCool, *From Unity to Pluralism*, p. 170 on Gilson's avoidance of the commentaries on Aristotle; see also James Collins, "Toward a Philosophically Ordered Thomism," pp. 280-299 in *Three Paths in Philosophy*, (Chicago: Regnery, 1962).

[9] The *Jerusalem Bible* tranlates "*Sum qui sum*" as "I Am who I Am"; some would read it, "I am that I am."

[10] John, Prologue, 1:1-4.

[11] The translation is by D. A. Patton, in *Essays Presented to Ernst Cassirer*, ed. R. Klibansky, (London: Oxford University Press, 1936), pp. 61-76; reprinted in *A Gilson Reader*, ed. A. C. Pegis, (Garden City, N.Y.: Doubleday, 1957), pp. 49-65; refs. are to this reprint.

investigation" (p. 60 in *Gilson Reader*), he comes to first philosophy and says:

> Thus Wisdom, or first philosophy, or metaphysics, lays down the guiding principles of all other sciences, and humanly depends on none of them, as the others study different modes of being, so it studies being itself . . .

Gilson is not suggesting there that philosophy depends on theology. He proceeds to speak of the special modes of mathematics, physics, biology, psychology and sociology. In this early work he says nothing about the academic role of theology. But Gilson is thinking of Christian philosophy: "Since the Christian universe is a creation of God, not of man, Christian philosophy instinctively looks on these problems from the point of view of the object." (p. 59)

This stress on objectivity, in a *Festschrift* honoring Kantian subjectivism, is ironical. To my mind, too, this objectivity is one of the essential characteristics of Thomistic philosophy. As Gilson insisted in his lectures on various kinds of epistemology: if your starting-points (*principia*) are restricted to mental data only, you will never be able to establish anything more than a world of subjective ideas. Of course Ernst Cassirer himself would not have been pleased with this stress on objectivity, nor would it be acceptable to Transcendental Thomists.

Gilson did not deny that there are other ways of philosophizing than the way of the Christian. As he wrote in 1957: "All love of wisdom so understood is philosophy. There are, therefore, a great many different ways of philosophizing, and many of them are unrelated to Christianity."[12] Yet in the same mature essay Gilson

---

[12] "What Is Christian Philosophy?" in *Gilson Reader*, p. 177; this essay was written for this *Reader*.

added (p. 182) that, "it should be obvious that all which is found in the *Summa Theologiae* is theological."

In the middle 1960's Gilson lectured in many centers in Canada and the United States. Since he practically always wrote his lectures in advance, typed copies are preserved in Gilson Archives (Pontifical Institute of Medieval Studies, Toronto) under the general title, "On the Renewal of Christian Philosophy." In one talk he argued that, "no genuine philosophical knowledge is possible without a minimum of metaphysical speculation."[13] And he added a statement which clearly shows that Gilson did not think that Thomistic first philosophy could ever evolve:

> There is a metaphysical knowledge whose truth is a unchangeable as the structure of the human mind . . . The contrary view is the popular one: it is now considered evident--which to Plato would have been a scandal--that Becoming, not Being is the core of reality.[14]

Later on, in 1971, Gilson published a book on evolutionary philosophy[15] which shows the philosophy of his maturity at work. Its fourth chapter, entitled "The Constants of Biophilosophy," offers many views that are typical. Gilson's basic claim is that Charles Darwin was a careful scientist who showed that species of living things do change--but the general philosophy of evolution was never proposed by Darwin. It is the

---

[13] See Shook, *E. Gilson*, p. 374.

[14] Shook comments on the same page: "In these papers Gilson articulated the respective roles of philosophy and theology as he would have liked the council [Vatican II] fathers to have done."

[15] *D'Aristote à Darwin et retour*, (Paris: Vrin, 1971); trans. by John Lyon, *From Aristotle to Darwin and Back Again*, (Notre Dame: Notre Dame University Press, 1984).

fabrication of later second-raters, like Herbert Spencer and Thomas Huxley.[16]

The main point that Gilson stresses, at this time, is that biological philosophy and biology must include a consideration of the ends of living things. All plants and animals are naturally ordered toward specific ends. The philosophy of living beings is necessarily teleological. Every kind of life species has an inbuilt *telos*, a perfection that is proper to it.

Now it seems to be that Gilson was an outstanding historian of medieval philosophy. I knew him well from the time of his first lectures at Toronto (1927) on late medieval theories of knowledge. I was in his first classes and seminars at the not-yet-Pontifical Institute of Medieval Studies. Again in the fall of 1937, when I finished my dissertation for the doctorate, I heard his lectures criticizing many recent types of Scholastic epistemology. In later years we met at various international congresses of philosophers. Our mature relations were cordial but I was never as close to Gilson as my classmate in graduate studies, Anton Pegis.

It amuses me to be classified as a "Gilsonian" in philosophy, for I differed from his views on many points. I still think that his Bonaventurean interpretation of Augustine's theory of intellectual illumination was wrong. Moreover, I enjoyed the friendship and shared some of the views of scholars from the University of Louvain: with Gerald Phelan (who eventually directed my dissertation on *habitus* in Aquinas' metaphysics), with Leon Noel (who taught an impressive course at Toronto in the summer of 1929 on Aquinas' theory of knowledge). For more than a decade I worked with Msgr. Louis De Raeymaeker on the affairs of the World Union of Catholic Philosophical Societies. There were thirty-five of them. Younger Louvain philosophers,

---

[16] See the Lyon trans. pp. 73-74.

such as F. Van Steenberghen, H. Van Breda, and G. Verbeke, have been my friends over the years.[17]

The reason for these personal details lies in the fact that I have concentrated on Gilson here, rather than on Rousselot, Maréchal and Maritain, men whom I never knew personally. In my opinion Gilson tried to devise a discipline lying midway between philosophy and theology. It must now be evident that I do not think that such a study is possible. There are many different ways employed by Christians who work at philosophy. I am in agreement with my late colleague James Collins.[18] He felt, as I do, that it is wrong to fuse philosophy with theology.

This does not mean that I deny Gilson's claim that certain themes in the Old and New Testaments have stimulated later philosophical thinking. Even atheistic critics of creation doctrines have owed a great deal to the Book of Genesis. Thomas Aquinas wore different hats (as Molière would say) when he talked philosophy or theology. The fact that there are philosophical treatises in the *Summa Theologiae* compiled by Aquinas does not make them theological. He tried to show people studying theology that there are philosophical explanations that help one to understand sacred teachings. There are, for instance, strictly philosophical arguments for concluding that a supreme being does exist, and that He is the real authority on which their beliefs rest. This is not to deny that the majority of religious people may firmly believe in the existence of God, without ever studying philosophy or theology. As Aquinas stated this: "there is nothing to prevent a person who cannot grasp a demonstration from accepting, as a

---

[17] See for instance, Gérard Verbeke, "A Crisis of Individual Consciousness: Aquinas' View," *The Modern Schoolman*, 69, 3 & 4 (1992), 379-394.

[18] See particularly pp. 382-389, in Collins' *Three Paths*.

matter of faith, something that in itself is capable of being scientifically known and demonstrated.[19]

As I understand the matter, Gilson did not show that there has been an "evolution" of Thomistic philosophy, into a plurality of ways of thinking. What he continually pointed out, as a historian of medieval thought, is that there were, from the eleventh to the fourteenth century, several different ways of doing philosophy in Christian centers. And there were different ways of relating philosophy to theology. There are the pathways of Anselm, Bonaventure, Albert, Aquinas, Duns Scotus, William Ockham, and many others. All espoused some form of Christian scholasticism. But only one was a Thomist.

V. Why Thomistic Philosophy Is not Pluralistic

Thomistic philosophy is not, and never has been, identical with the whole of philosophy taught by Catholics. Nor is it the same as Scholastic philosophy. Basically it is the philosophy found in Thomas Aquinas' theological writings and in his commentaries on Aristotle, Pseudo-Dionysius, and Boethius.

Some major features of this philosophy of Thomas Aquinas are briefly stated here, without documentation, for there are simply key themes that seem valid to me after reading Aquinas for more than sixty years. Other people may have different things that they would stress.

To be (*esse*) is very different from whatness (*essentia*). Essences are universal: there are no individual essences (versus Nominalism and Suarezian metaphysics).

---

[19] *S.T.* I, 2, 2, ad 1m: Nihil tamen prohibet illud quod secundum se demonstrabile est et scibile, ab aliquo accipi ut credibile, qui demonstrationem non capit. (Ottawa ed. 1941, vol. I, 13a.)

Causal explanations require not only the concept of a productive source (efficient cause) but also material, formal and final causes.

Potency and act are useful factors for the analysis of coming into being, and of all sorts of operations.

The most distinctive feature of human knowledge is its objectivity (versus transcendentalism). This means that what guarantees the truth of the philosopher's judgments is the fact that existing things thrust certain items of knowledge upon his consciousness. Through sense perception, cogitative judgment on such percepts, and intellectual judgment based on such presentations, humans know the world about them, their own persons, and the existence and nature of other persons. Understanding of the universal natures of things and persons is objective, in the sense that it is not a fabrication of inward-looking consciousness.

Final causality, the notion of end-directedness in all kinds of finite beings, is a valuable explanatory principle in practical philosophy.

In the practical order of human thinking, looking to personal voluntary action and omission (ethics, artistic operations, socio-political activities), the judging is done by the person using standards or laws grasped initially by intellectual synderesis (a *habitus* of the understanding that is inborn to intuit such rules as the principle of non-contradiction, or the axiom that one should do only what is judged to be good and right, for self and others). Such general rules and conclusions issuing from them are applied to personal individual problems by prudential judgments that involve sense perception of concrete situations, intellectual decisions and volitional (will) action.

There is a radical difference between ethics (general propositions about universal types of human activity) and prudential discourse (looking to personal, individual problems, whether to act or not, whether to do this act or another act, under these concrete circumstances). There is no science or philosophy that solves such personal problems.

Since the circumstances of human living vary greatly from century to century, changes in human philosophizing are needed in both the speculative and the practical order. I do not think these changes modify our general understanding of natural things in their universal characteristics. It is quite possible that the species, or specific natures of some plants and animals, and even inanimate objects, have changed during long periods of time. But, as I see them, the men and women that we read about in ancient literature and history were the same kind of people and had the same sort of moral problems that we face today.

If, under the name of neo-thomism, one calls the kind of new philosophy that uses the initial principles found in Thomas Aquinas' thinking "thomistic"--I have no objection to that. It is necessary to adapt to present-day circumstances. There must be some diversified conclusions, speculative and practical, that were not known to Aquinas. The pluralism carefully indicated by McCool does occur in such neo-thomism. However, I do not think that the general positions that I have attributed to Aquinas admit of pluralization.

THE UNITY OF THOMISTIC EXPERIENCE--
A GILSONIAN REJOINDER TO
GERALD MCCOOL, S.J.

Peter A. Redpath

After having recently re-read parts of Father Gerald McCool's intriguing and provocative "history" of the non-development of Neo-Thomism,[1] I could not help but wonder what would have been Etienne Gilson's reaction to the amiable Jesuit's depiction of Gilson's major role in "undermining the Neo-Thomistic movement."[2] In fact, the charge seems to me to be so surreal that I cannot avoid picturing Gilson, in the figure of Socrates in Plato's *Apology*, reacting to the description given of him by his eloquent accuser and stating that, based upon it, he "almost forgot who he was."[3] In addition, I also cannot avoid imagining Gilson and Father McCool respectively as Père Malebranche and Bishop Berkeley in an incident retold by Gilson in his *The Unity of Philosophical Experience*:

> If young Berkeley did use such an argument, which I have borrowed from his later criticism of Malebranche, the account given by Stock of their interview is not entirely lacking in probability: "In the heat of the disputation," says Stock, "he [Malebranche] raised his voice so high, and gave way so freely to the natural impetuosity of a man of parts, and a Frenchman, that he brought on himself a violent increase of his disorder, which carried him off a few days after." If the story is true, it is a good one; if it is not true, it is better than true, for it

---

[1] Gerald A. McCool, S.J., *From Unity to Pluralism: The Internal Evolution of Thomism* (New York: Fordham University Press, 1989).

[2] *Ibid.*, p. 197.

[3] Plato, *Apology*, 17A.

> should have happened. No wonder then, that DeQuincey inserted it in his famous Essay on *Murder as One of the Fine Arts*. What a murder case, indeed: "Murder by Metaphysics!"[4]

Just as Gilson says about Stock's account of the interchange between Berkeley and Malebranche, if the story "is not true, it is better than true, for it should have happened," so, too, I believe is my fictional exchange between Father McCool and Etienne Gilson. For while it is not true that any such exchange ever did happen, my fantasy is better than true (if that is possible) for it should have happened. Perhaps, however, in reality it is better that it did not happen; for, as I envision the outcome, Gilson, as "a Frenchman" and "a man of parts," upon hearing of Father McCool's analysis of his own role in the murder of Neo-Thomism would *not* have "brought on himself a violent increase of his disorder" and would *not* have been "carried off a few days after"; but immediately, "as a Frenchman" and "a man of parts," he would have sought to get his hands on his interviewer in a not-so-metaphysical way.

Indeed, it is sad that Gilson is not alive to respond in person to Father McCool's charges, especially since a number of the things which McCool says, viz., about Gilson, Gilson's interpretations of St. Thomas, Neo-Thomism, and the papal encyclical *Aeterni Patris*, all appear to me to be oversimplifications, distortions, and historical inaccuracies. As perhaps the greatest twentieth-century historian of medieval Catholic thought, I think that Gilson would have been particularly galled by Father McCool's conclusion that Gilson's own work had seriously undermined the Neo-Thomistic movement--especially since Father McCool's interpretation of Gilson seems to me to be based upon a kind of Protagorean Thomism which colors both the

---

[4] Etienne Gilson, *Unity of Philosophical Experience* (New York: Charles Scribner's Sons, 1965), p. 195.

good Father's own historical research and his own reading of texts.

In a sense, what Father McCool has done to Gilson has been to "murder," if not him, at least his reputation, both historically and otherwise. It seems to me that such an unjust act warrants some defense on the part of the victim, which is what, in whatever feeble fashion I might be capable of so doing, I will attempt to achieve in this essay.

To show the many ways in which I think Father McCool has wrongly proclaimed the death of the Neo-Thomistic movement, and to correct his mistaken evaluation of Gilson's supposed role in its demise, let me begin by outlining Father McCool's account of the whole misbegotten affair--beginning with the birth of Neo-Thomism during the pontificate of Leo XIII--in particular, with respect to its conception in Pope Leo's encyclical *Aeterni Patris*.

As McCool sees it, within this encyclical Leo set forth a program for the development of Neo-Scholasticism as a "common doctrinal tradition," a philosophy common to all the Scholastic doctors. Moreover, Gilson had been able to agree with the encyclical (which Gilson had honestly admitted he had not read "until many years after he had worked out his own conception of Christian philosophy") because he had "understood the encyclical to be saying what in fact it never says: that to be a Thomist is to adhere with absolute fidelity to the way of philosophizing St. Thomas employed in his theological works."[5] In addition, because, through his own historical research, Gilson had eventually proven that "there was no common doctrinal tradition" "in Patristic and medieval philosophy and theology," in McCool's view, *in reality*, and despite the fact that "he always referred to the encyclical in tones of agreement and high praise,"

---

[5] McCool, *From Unity to Pluralism*, p. 197.

"Gilson could not accept the validity of the program for Neo-Scholasticism's development set forth in *Aeterni Patris*."[6]

By saying that "Gilson could not accept the validity of the program for Neo-Scholasticism's development set forth in *Aeterni Patris*" McCool does not seem to mean that Gilson had engaged in some sort of open rebellion and criticism of this famous encyclical of Leo XIII. No, what Father McCool has in mind seems to be that Gilson's own research had led him to reach certain conclusions regarding the nature of Christian philosophy which contradicted the program for Neo-Scholasticism's development as set forth by Pope Leo in *Aeterni Patris*. In a sense, as McCool sees it, Gilson had done this unwittingly by discovering that there was no common doctrinal tradition to medieval Scholasticism, Neo-Thomism, Neo-Scholasticism, or Christian philosophy. For "due to his research, the accepted belief that Thomism incorporates a common doctrinal tradition inherited from the medieval Doctors through the great post-Tridentine Scholastics lost its historical foundation."

Thus, through the very sort of logical necessity which Gilson himself had so often associated with the "unity" of "philosophical experience," McCool thinks Gilson's own research leads to the logical conclusion that the Neo-Thomistic movement not only no longer exists, but, even more paradoxically, that it never began. For how could it have begun when the very conditions for the possiblity of its existence were, in and of themselves, non-existent? As Father McCool puts it:

> Due to his research, the accepted belief that Thomism incorporates a common doctrinal tradition inherited from the medieval Doctors through the post-Tridentine Scholastics lost its historical foundation. The effect of its loss on the Neo-Thomistic

---

[6] *Ibid.*, pp. 196-97.

> movement was severe. As Gilson himself observed, if his criteria for determining Thomism are the right ones, there can still be Thomists but there can be no Neo-Thomists. For the authentic disciple of the Angelic Doctor progress can consist only in his ever-deepening understanding of what St. Thomas himself has written.[7]

Of course, since, as Father McCool sees it, Gilson showed that the Neo-Thomistic movement had "lost its historical foundation," it makes no sense for Father McCool to end his text with the rather cavalier claim: "The history of the modern Neo-Thomistic movement, whose *magna carta* was *Aeterni Patris*, reached its end at the Second Vatican Council."[8] Clearly, if the Neo-Thomistic movement really had been shown by Gilson to have lost its historical foundation, the history of the Neo-Thomistic movement certainly could not have ended with the Second Vatican Council. For the "movement" had been no movement at all. Consequently, it could not have had any history in the first place!

Father McCool's proclamation of the death, or complete non-existence, of the Neo-Thomistic movement, however, seems to me to be premature. For his conclusion, at best, is based upon a number of questionable and dubious assumptions on his part regarding the "program" of *Aeterni Patris*, the nature of Scholasticism and Neo Scholasticism, of Thomism and Neo-Thomism, and of the nature and impact of Gilson's research.

In addition, the very fact that I am calling his thesis into question should give readers pause to wonder about the validity of his conclusion. Why is this so? Because if there were anyone who reasonably might be expected to support Father McCool's position regarding the non-existence of a Neo-Thomistic movement, it

---

[7] *Loc. cit.*

[8] *Ibid.*, p. 230.

would be me. For the very sort of conclusion which Father McCool has drawn seems to follow with logical necessity from some of my own research. For example, as far back as 1987, in an article entitled "Romance of Wisdom: The Friendship between Jacques Maritain and St. Thomas Aquinas," I criticized the notion of Christian philosophy defended by both Gilson and Maritain. Regarding medieval Christian philosophy in general and Gilson's concept of Christian philosophy in particular, I said:

> If such be the case, if no master in the thirteenth-century would even have considered himself as having a philosophy, even a Christian one, if the Scholastic theologian of the Middle Ages was normally not described as a philosopher but as a philosophizing theologian or as a "philosophizer," if the term "philosopher" was usually restricted to pagan thinkers, and if the thirteenth-century theologians do not seem explicitly to have considered the possibility of a person simultaneously being a philosopher and a Christian, how then can it be considered fitting to describe the spirit of "philosophy" of this period as Christian philosophy? Indeed, the spirit of medieval "philosophy" is not a philosophical spirit at all. Rather, it is a theological one. There is no philosophy as such in the Christian Middle Ages at all.[9]

Furthermore, Father McCool's conclusion would also seem to have the support of no less of an authority than the great contemporary Jesuit historian of philosophy F.C. Copleston. For, commenting upon my criticism of Maritain's notion of Christian philosophy, with which,

---

[9] Peter Redpath, "Romance of Wisdom: The Friendship between Jacques Maritain and Saint Thomas Aquinas," in *Understanding Maritain: Philosopher and Friend* (Macon, Mercer University Press, 1987), p. 102.

for the most part, Gilson substantially concurred, Father Copleston has recently observed:

> ...when discussing Maritain's idea of "Christian philosophy" and his distinction between the "nature" and "states" of philosophy, Professor Redpath aptly remarks that "apart from its state philosophy has no existence at all" (p. 110). For good measure, Redpath describes Maritain as "a Scholastic theologian" (p. 113), though this description is not intended to be derogatory.[10]

Clearly, given my own position on the notion of Christian philosophy, and its seemingly positive reception from no less than Father McCool's esteemed Jesuit colleague, some readers might wonder how I could possibly disagree with McCool's conclusion about the non-history of Neo-Thomism. Indeed, they might even raise the issue of my previous statements regarding the nature of Christian philosophy as a foil against any criticism I might now seek to raise against Father McCool.

Nevertheless, I think his view of Christian philosophy and my view are in extreme opposition (but just what is the nature of that opposition is not something upon which I can elaborate in this article),[11] and, consequently, I do not find any problem criticizing Father McCool's conclusions and supporting premises. More precisely, I think his reading of *Aeterni Patris*, at times, not only tortures the meaning of the text but also projects into the encyclical meanings which are not supported by the text. In addition, I think he distorts the natures of Thomism and Neo-Thomism, of Scholasticism and Neo-Scholasticism, and of the impact

---

[10] F.C. Copleston, review of *Understanding Maritain: Philosopher and Friend*, in *Heythrop Journal*, 32 (1991), 444.

[11] I am in the process of writing a book which, in part, will spell out more precisely my notion of "Christian philosophy."

of Gilson's research upon the Neo-Thomistic movement. For, in my opinion, that research does not show that Neo-Thomism lacks a "common doctrinal tradition"; and, even if it did, this would be of no import. For Father McCool's assertion that the lack of a common doctrinal tradition would, in some way, undermine the historical roots of Neo-Thomism, because it would undermine the "program" of Leo XIII's encyclical *Aeterni Patris*, has no foundation in reality; its only foundation lies in Father McCool's cabalistic reading of this papal letter which is generally unsupported by the text itself.

To support some of the claims which I have just made in the above paragraph, I think it should be noted, first of all, that, instead of beginning his own work, as one might expect, with a detailed analysis of *Aeterni Patris* itself, Father McCool begins his first chapter with a review of the nineteenth-century Neo-Scholastic movement which influenced Leo's thinking. He spends about five pages summarizing Leo's views of "Christian philosophy" and of his conceptions of philosophy and theology and over twenty pages dealing with the surrounding nineteenth and early twentieth century circumstances within which Leo's encyclical developed.[12] The following passage more or less summarizes Father McCool's analysis of Leo's views about philosophy, theology, and Neo-Thomism:

> In the encyclical's description of theology, modeled on timeless Aristotelian science which moves from first principles to certain conclusions, we can recognize the type of "conclusion-theology" that, as Johannes Beumer has shown, dominated the great deal of speculative theology after Vatican I. Aristotelian science has no place for history, and the Neo-Scholastics who took it as their model for theology lacked the awareness of history's importance for

---

[12] See McCool, *From Unity to Pluralism*, pp. 5-38.

> theology which their German rivals possessed. In their opinion the sound philosophical theology that enabled St. Thomas to integrate the essence of the Church's Patristic tradition into his theology and to hand it down to posterity would enable the modern scholastic theologians to integrate the data of modern scholarship into their theology. Problems of historical development, hermeneutics, or diverse conceptual frameworks did not trouble them.
>
> The same lack of historical perspective affected their understanding of the "wisdom of St. Thomas." *Aeterni Patris* gives the impression that all the Scholastic Doctors had the same philosophy and theology. St. Thomas was the best of them, but there is no difference between the philosophy that structures his theology and the philosophy of St. Bonaventure. Nor does the the encyclical show any awareness of historical development within Scholasticism itself, or of the notable difference between the scholastics of the sixteenth and seventeenth centuries and the Scholastic doctors of the Middle Ages. Cajetan and Pope Sixtus V are looked on as defenders of Thomas' own philosophy. No sign is given that the diversity among contemporary Thomists, Suarezians, or even Scotists was of great philosophical significance. Their differences were on points of lesser moment, and, in Leo's eyes, all apparently might be called worthy disciples of Thomas and the great Scholastic Doctors.[13]

It seems clear Father McCool thinks that, in his encyclical, Pope Leo was advocating a return to a philosophy and theology "common to the Scholastic Doctors whose finest exponent was St. Thomas."[14] Indeed, that Leo was advocating such a position is so evident to Father McCool that, as I asserted above, he even went so far as to criticize Gilson for understanding "the encyclical to be saying what in fact it never says:

---

[13] *Ibid.*, p. 11.

[14] *Ibid.*, p. 8.

that to be a Thomist is to adhere with absolute fidelity to the way of philosophizing St. Thomas employed in his theological works."[15] Instead, as Father McCool sees it, what the encyclical does say is that to be a Thomist is not to adhere to *a way of philosophizing* employed in theological works, but to adhere to a philosophy, to a universal philosophy and a theology, for that matter, modeled after non-historical Aristotelian science.

Now it seems to me that before making such exaggerated claims Father McCool would have done well to read the introductory section entitled "How to Read the Encyclicals" in Gilson's edition of several papal letters, including *Aeterni Patris*, entitled *The Church Speaks to the Modern World.* For in that work Gilson makes the following sage observations:

> When a Pope writes such a document, he does so in the full awareness of his spiritual responsibility. He knows very well that each and every sentence, word, noun, epithet, verb, and adverb found in his written text is going to be weighed, searched, and submitted to the most careful scrutiny by a crowd of countless readers scattered over the surface of the earth. And not only this, but the same anxious study of his pronouncements will be carried on by still many more readers, including his own successors, for generation after generation.
>
> This thought should be an invitation to approach these texts in a spirit of reverence and intellectual modesty. The teaching of the encyclicals should not be made either broader in scope or more narrow than it is. Dealing as it does with a restatement of the Catholic faith as well as with its applications to definite problems, this teaching must be understood as given. Only a Pope has authority to complete the teaching of one of his own encyclicals as well as that of the encyclicals of other Popes, since only a Pope

---

[15] See note 4 above.

> has authority to write and to publish such a document.
>
> Another rule to observe is not to yield to the temptation of "improving" the doctrine of the Popes. In commenting [on] the encyclicals, many consider it their duty either to broaden the meaning of the doctrine or, more often still, to tone it down in the hope that, with a few minor adjustments, it will become more palatable to the modern mind. Generally speaking, this well-intentioned desire to better the teaching of the Popes springs from misinterpretation. When it seems to us that an encyclical cannot possibly say what it says, the first thing to do is to make a new effort to understand exactly what it does say. Most of the time it will then be seen that we had missed, not only the meaning of its terms, but also the general intention of the document and even, in some cases, the very nature of the problem under discussion.[16]

Specifically, in the case of Father McCool, I think that Gilson's observations are relevant for their advocacy of moderation in interpreting the meaning of a papal encyclical. I am not referring to Gilson here in order to accuse Father McCool of intellectual immodesty or of a desire to "improve" the doctrine of Pope Leo, however. My only point is simply that Father McCool's interpretation of Leo suffers from the weaknesses of excess and defect. For clearly he reads into the words of the document notions which are not justified by the text.

For example, the text of *Aeterni Patris* carefully avoids giving any definition whatsoever of "philosophy" or of "Christian philosophy," and what it stresses is not philosophy as a common doctrine but the "use" of philosophy in relationship to theology and in relationship

---

[16] Etienne Gilson, *The Church Speaks to the Modern World* (Garden City, New York: Doubleday and Company, 1954), pp. 21-22.

to the Catholic faith. Indeed, as Joseph Owens rightly observes:

> The concern of the document is solely with the use of philosophy in general for serving the interests of Christian faith as philosophy had actually done in the past.
>
> The question of what constitutes a Christian philosophy is accordingly not a question faced by *Aeterni Patris*.[17]

In fact, as Father Owens notes, the phrase "Christian philosophy" does not even occur in the text of the encyclical letter, although it was later used by the "Pope himself as the document's title on the first anniversary of its publication."[18] In addition, according to Owens, quite in opposition to Father McCool, as used in the encyclical, the term "philosophy" is not being used to refer to a common doctrine of timeless Aristotelian science held by Scholastics or by anyone else for that matter:

> By "philosophy," the encyclical means what we today would call the whole philosophical enterprise. It sketches the history of philosophy only in the western tradition, but says nothing that would interfere with our present extension of interest to Persian, Hindu, and Chinese philosophies. What it has in mind is philosophy in general. It is concerned with "the right use of philosophy" understood in this global fashion. In that use philosophy's own distinctive starting points and methods are guaranteed, yet the right to use them in total independence of divine revelation is not sanctioned. What is

---

[17] Joseph Owens, *Towards a Christian Philosophy* (Washington, D.C.: The Catholic University of America Press, 1990), p. 73

[18] *Ibid*., p. 63.

envisaged is a *philosophical genus* (p. 100.32) that opens a clear and facile way towards faith.[19]

It is interesting to note, furthermore, that in his own citation of a key passage from Pope Leo's letter in which the Pope articulates his idea of the best way to use philosophy, Father McCool quotes the Pope as saying: "*Aeterni Patris* considers that those who 'combine the pursuit of philosophy with dutiful obedience to the Christian faith' are the best philosophers."[20] Yet in his own footnote to this passage, McCool calls into question the accuracy of his very own translation in favor of the one used by Gilson!: "'The best philosophers' is Maritain's translation. Gilson more accurately translates the phrase as 'philosophizing in the best possible way'..."[21] Indeed, Gilson's translation is better not only because it agrees with the Latin original but also because the chief concern of the Pope was not with promoting a program based upon a common doctrinal tradition of conclusions in philosophy or theology but with a way of using philosophy in relationship to the Christian faith, with a way of doing philosophy and theology after the fashion of the way St. Thomas did them.

In his encyclical Pope Leo had identified two weaknesses in particular regarding the contemporary status of "philosophy." One was the divorce of the principles of "modern philosophy" from a real foundation in the nature of sensible things, and the other was the divorce of modern philosophy from theology. Regarding both of these points Leo gave the reasons for his support of the method of "philosophizing" used by St. Thomas and the superiority of "Scholastic philosophy." When one reads these reasons objectively, it is clear that what Leo was advocating was not a

---

[19] *Ibid.*, p. 67.

[20] McCool, *From Unity to Pluralism*, p. 9.

[21] *Ibid.*, p. 36.

"conclusion-theology" or any "common doctrinal tradition" in the sense of commonly accepted conclusions which he identified with what Father McCool calls "timeless Aristotleian science." No, what Leo was recommending was the classical method of philosophical realism and the proper way of relating such philosophical realism to the Catholic faith and to revealed theology. As Leo saw it, the Scholastics preserved the wisdom of the ancient Greeks by not losing realism in the method of deriving their first principles, and St. Thomas had perfected the method of using this "philosophical-principle-realism" as a handmaiden for theology.

Consequently, in praise of the "philosophy of the ancients," that is, in praise of philosophical realism (which Gilson would later call "methodical realism"[22]), Leo says:

> And, assuredly, the God of all goodness, in all that pertains to divine things, has not only manifested by the light of faith those things which human intelligence could not attain of itself, but others, also, not altogether unattainable by reason, that by the help of divine authority they may be made known to all at once without the admixture of error. Hence it is that certain truths which were either divinely proposed for belief, or were bound by the closest chains to the doctrine of faith, were discovered by pagan sages with nothing but natural reason to guide them, were demonstrated and proved by becoming arguments. For, as the Apostle says, the invisible things of Him, from the creation of the world, are clearly seen, being understood by the things that are made: His eternal power also and divinity; and the Gentiles who have

---

[22] See Etienne Gilson, *Methodical Realism*, trans. Philip Trower (Front Royal, Virginia: Christendom Press, 1990); and *Thomist Realism and the Critique of Knowledge*, trans. Mark A. Wauck (San Francisco: Ignatius Press, 1986).

> not the Law show, nevertheless, the work of the Law written in their hearts.[23]

In other words, it is precisely because the ancient philosophers were realists, precisely because, that is, they had derived their philosophical first principles from their knowledge of the natures of physical things that, for Pope Leo, these philosophers were able to prove many "invisible things" about God by their natural reason alone. In fact, it is because St. Thomas follows this same procedure of deriving his philosophical first principles from his knowledge of the being of physical things that he, too, and others as well, in Leo's mind, have been able, over the centuries to build upon the foundation laid by the ancient Greeks:

> Moreover, the Angelic Doctor pushed his philosophic inquiry into the reasons and principles of things, which because they are most comprehensive and contain in their bosom, so to say, the seeds of almost infinite truths, were to be unfolded in good time by later masters and with a goodly yield.[24]

Furthermore, in the way he used philosophy, St. Thomas was careful always to distinguish his principles of natural reason from his principles which surpass the capacity of reason to grasp. By so doing, Pope Leo says, St. Thomas was able happily to associate reason and faith:

> Again, clearly distinguishing, as is fitting, reason from faith, while happily associating the one with the other, he both preserved the rights and had regard for the dignity of each; so much so, indeed, that reason, borne on the wings of Thomas to its height, can scarcely rise higher, while faith could scarcely expect

---

[23] Pope Leo XIII, *Aeterni Patris*, in Gilson, *The Church Speaks to the Modern World*, p. 34.

[24] *Ibid.*, p. 44.

> more or stronger aids from reason than those which she has already obtained through Thomas.[25]

As Leo sees it, this happy marriage of classical philosophical realism and Christian faith was continued by other Scholastics coming after St. Thomas until a messy divorce occurred in the sixteenth century which has continued therafter:

> For it pleased the struggling innovators of the sixteenth century to philosophize without any respect for faith, the power of inventing in accordance with his own pleasure and bent being asked and given in turn by each one. Hence it is natural that systems of philosophy multiplied beyond measure, and conclusions differing and clashing one with another arose about those matters even which are most important in human knowledge. From a mass of conclusions men often come to wavering and doubt; and who knows not how easily the mind slips from doubt to error? But, as men are apt to follow the leads given them, this new pursuit seems to have caught the souls of certain Catholic philosophers, who, throwing aside the patrimony of ancient wisdom, chose rather to build up a new edifice than to strengthen and complete the old by aid of the new--ill-advisedly, in sooth, and not without detriment to the sciences. For, a multiform system of this kind, which depends on the authority and choice of any professor, has a foundation open to change, and consequently gives us a philosophy not firm, and stable, and robust like that of the old, but tottering and feeble. And if, perchance, it sometimes finds itself scarcely equal to sustain the shock of its foes, it should recognize that the cause and blame lie in itself. In saying this we have no intention of discountenancing the learned and able men who bring their industry and erudition, and, what is more, the wealth of new discoveries, to the service of

---

25 *Loc.cit.*

> philosophy; for, of course, we understand that this tends to the development of learning. But one should be very careful lest all of his chief labor be exhausted in these pursuits and in mere erudition. And the same thing is true of sacred theology, which, indeed, may be assisted and illustrated by all kinds of erudition, though it is absolutely necessary to approach it in the grave manner of the Scholastics, in order that, the forces of revelation and reason being united in it, it may continue to be "the invincible bulwark of the faith."[26]

As the above passage indicates, the chief purpose of Leo's encyclical is to end the divorce between Catholic theology and classical philosophical realism. Without such a restoration, Leo thinks that not only will revealed theology suffer but so also will all the human arts, in particular the liberal arts, as well as the physical sciences. Consequently, toward the end of the encyclical he summarizes the universal benefits which he envisions his program could bring:

> In short, all studies ought to find hope of advancement and promise of assistance in this restoration of philosophic discipline which We have proposed. The arts were wont to draw from philosophy, as from a wise mistress, sound judgment and right method, and from it also, their spirit, as from the common fount of life. When philosophy stood stainless in honor and wise in judgment, then, as facts and constant experience showed, the liberal arts flourished as never before or since; but, neglected and almost blotted out, they lay prone, since philosophy began to lean to error and to join hands with folly. Nor will the physical sciences themselves, which are now in such great repute, and by the renown of so many inventions draw such universal admiration to themselves, suffer detriment, but find very great assistance in the restoration of the

---

[26] *Ibid.*, pp. 46-47.

> ancient philosophy. For, the investigation of facts and the contemplation of nature is not alone sufficient for their profitable exercise and advance; but, when facts have been established, it is necessary to rise and apply ourselves to the study of the nature of corporeal things, to inquire into the laws which govern them and the principles whence their order and varied unity and mutual attraction in diversity arise. To such investigations it is wonderful what force and light and aid the Scholastic philosophy, if judiciously taught, would bring.[27]

It seems to me it should be evident by this time that the "Scholastic philosophy" which Pope Leo XIII envisions reviving is not some sort of "timeless Aristotleian science" which has no respect for history (whatever that is--certainly it has nothing to do with either the way Aristotle or St. Thomas did science[28]), or some sort of "conclusion-philosophy" or "conclusion-theology." No, what Leo has in mind was the sort of Scholasticism which Gilson describes in his *Methodical Realism*:

> Scholasticism is a self-conscious, a considered, a willed realism. It is not based on a solution to a problem posed by idealism for the very reason that the givens in such a problem necessarily imply the solutions of that same idealism. In other terms, surprising as a thesis like this at first-hand appears, scholastic realism is not a function of the problem of knowledge--the contrary is rather the case--but the real is put there as something distinct from thought, the

---

[27] *Ibid.*, p. 49.

[28] Just what Father McCool means by "timeless Aristotelian science" is difficult to determine. First of all, the science of Aristotle is not the science of St. Thomas, and neither Aristotle nor St. Thomas was oblivious to the impact of the research of others upon their own work. While neither had the advantages of contemporary research tools, nonetheless, both were aware of, and put into practice, the benefits of historical research in their work. To claim otherwise is itself a distortion of history.

> *esse* is put there as distinct from *percipi*, and this because of a definite idea of what philosophy is, and even as a contradiction of the very possibility of philosophy. It is a methodic realism....[29]

In addition, he says:

> With a sure instinct for the right way, the Greeks took resolutely the road of realism, and the scholastics did so too because it led somewhere. Descartes tried a different road. Once walking it, he saw no obvious reason for not going on. We today, however, know this way leads nowhere, and so we have to leave it....[30]

And, finally, dispelling any doubt that Gilson conceives of Scholasticism as some sort of "timeless Aristotelian science" or "conclusion-theology" making metaphysical pronouncements from on high--whether these be Aristotelian heights or Cartesian heights--he states:

> Scholastic realism is not based on metaphysical reasoning. If this reasoning starts from God, it has to fail because of the impossibility of deducing the contingent from the necessary. If it starts from thought in Descartes' sense, it also fails, but for a different reason. Between one contingent being and another there is always a metaphysical rupture because of the analogy of being: if the being from which one departs is heterogeneous to the other, one never reaches the other because the being of that other will never be anything more than a representation of itself. Thus the only solution left to admit, even as experience suggests, is that the subject does not find its object in an analysis of knowledge of it, but it

---

29 In L.K. Shook, *Etienne Gilson* (Toronto: Pontifical Institute of Mediaeval Studies, 1984), p. 220.

30 *Loc. cit.*

> finds its knowledge and itself too, in an analysis of its object....[31]

Indeed, as Joseph Owens again correctly observes, the principles of philosophy which Christian theology puts to use must be derived from naturally accessible objects, that is, from sensible things:

> But what *Aeterni Patris* (p. 103.17) understands as philosophy in service of revealed doctrine is a discipline that has its own method, its own starting points, and its own reasoning processes. It arrives at its conclusions in a way other than that of faith (p. 103.14). In consequence it can receive neither method nor content from faith or from theology. Its content must be derived entirely from naturally accessible objects and, in the instance of Aquinas, from sensible things. It cannot allow even any mixing of a little theology into philosophy. The philosophy that can be of service to the faith has to be pure philosophy.[32]

Thus, if *Aeterni Patris* is promoting any sort of "common doctrinal tradition," this "common doctrinal tradition" is not one of a "conclusion-theology" or of a "conclusion-philosophy." Rather, it is one of a *habitus* of reasoning involving philosophical realism working in the service of revealed theology. What Leo has in mind is not a set of conclusions, or a body of knowledge, passed on from one generation to the next.[33] Thomism

---

[31] *Ibid.*, pp. 221-222.

[32] Joseph Owens, *Towards a Christian Philosophy*, p. 74.

[33] As Father Armand A. Maurer has correctly pointed out, for St. Thomas, properly speaking, science is a *habitus* of the intellect and not a "body of knowledge." See Armand A. Maurer, "The Unity of a Science: St. Thomas and the Nominalists," in *St. Thomas Aquinas, 1274-1974, Commemorative Studies*, II vols., editor-in-chief Armand A. Maurer (Toronrto: Pontifical Institute of Mediaeval Studies, 1974), II, pp. 269-291.

is a habit of reasoning, and those who habitually reason according to this minimal doctrinal tradition merit the title of Neo-Thomist even if they do not share all of the conclusions of St. Thomas himself.

Consequently, even those Neo-Scholastics who, like John of St. Thomas and Cajetan, might not correctly have understood St. Thomas's metaphysics of being, are justifiably described as Neo-Thomists and Neo-Scholastics. The brand of "Thomism" which Father McCool is advocating in his "pluralism," however, is without justification because it never puts philosophy to use, and it never puts philosophy to use because it involves reasoning without realism; and reasoning without realism is fantasy, not philosophy. The Thomism he offers, in short, is no philosophy and no theology at all. What it is, in fact, is poetry and allegory posing as philosophy and theology.

Were contemporary Thomists to accept Father McCool's analysis of modern Thomism, then, the future of Neo-Thomism and of Catholic philosophy and theology would be easy to predict. For it would have no future at all; it would be truly dead. Indeed, to recognize quite clearly that what I am saying is true, all one has to do is to reconsider a couple of prophetic passages written by Pope Leo XIII, and cited above,[34] about the future of philosophy when divorced from realism and to substitute in various places the word "Neo-Thomism" and other suitable words for other terms used by Leo. Hopefully, the wisdom of Leo XIII will convince the reader better than I ever could what sort of future lies in store for Catholic education if it falls prey to the adoption of the pluralism and poetized Aquikantianism[35] of Father McCool. If contemporary Thomists follow McCool's lead, then, echoing the words of Pope Leo, when they find themselves unequal to withstand the attacks of their

---

[34] See above notes 26 and 25.

[35] I believe this term was coined by Stanley Jaki.

enemies, they should recognize that the cause for their weakness will lie within themselves; for they will, indeed, have been lacking in Thomistic experience.

GILSON AND *AETERNI PATRIS*

Armand A. Maurer, C.S.B.

In his book *From Unity to Pluralism* Gerald McCool traces the history of scholasticism from the nineteenth-century notion of a unitary system of philosophy and theology to its fragmentation in the twentieth century into a plurality of opposing Thomisms.[1] He devotes two chapters (7 and 8) to Etienne Gilson as one of the main contributors to this breakdown. Through his research into medieval philosophy, Gilson showed that the notion of a common philosophical tradition in the Middle Ages is an illusion. Bonaventure, Thomas Aquinas, Duns Scotus, among other medieval philosophers, professed highly original doctrines. They did not share a common tradition among themselves or with the Fathers of the Church, such as Augustine.[2] Thus, according to McCool, Gilson shattered the dream of a common Christian philosophy that began with the Church Fathers, was systematized by the medieval scholastics, and developed by modern Thomists.

While Gilson was destroying the neoscholastic notion of a common Christian philosophical tradition, at the same time-- McCool contends--he was undermining Leo XIII's program in his encyclical *Aeterni Patris*,

---

[1] Gerald A. McCool, *From Unity to Pluralism: The Internal Evolution of Thomism* (New York: Fordham University Press, 1989). For the background of nineteenth-century scholasticism and Thomism see the same author: *Catholic Theology in the Nineteenth Century: The Quest for a Unitary Method* (New York: The Seabury Press, 1977); Thomas J. A. Hartley, *Thomistic Revival and the Modernist Era* (Toronto: University of St. Michael's College, 1971).

[2] G. McCool, *From Unity to Pluralism*, p. 196.

which was the Church's official endorsement of the neoscholastic movement.[3] The encyclical reflects the vision of Kleutgen, Liberatore, and the other pioneers of the movement, who taught the existence of a traditional philosophy in the Church and sought to restore the doctrine of St. Thomas as its finest expression. Since Gilson proved that there was no common scholastic doctrine, he could hardly accept the Pope's program. And yet Gilson highly praised the encyclical, voicing his agreement with it and extolling the Pope as one of the great Christian philosophers. In fact, according to McCool, "Gilson was able to agree with *Aeterni Patris* because he understood the encyclical to be saying what in fact it never says: that to be a Thomist is to adhere with absolute fidelity to the way of philosophizing St. Thomas employed in his theological works." [4]

It is possible, of course, that Gilson misread the encyclical, but it seems unlikely in view of his known ability to read documents accurately. Did he believe he found in the encyclical what McCool says he did? If not this, what was it that Gilson discovered there that aroused his intense admiration and prompted his enthusiastic agreement with it?

Gilson was well aware of the history of nineteenth-century scholasticism and the role Leo XIII

---

[3] The encyclical was issued August 4, 1897. It is usually entitled "On Christian Philosophy." This title does not appear in the encyclical but it was used by the Pope himself a year later in an Apostolic Letter. See G. Van Riet, "Le titre de l'encyclique "'Aeterni Patris,'" *Revue philosophique de Louvain*, 80 (1982), 35-63. On the encyclical see G. McCool, *ibid.*, pp. 5-38. *One Hundred Years of Thomnism: Aeterni Patris and Afterwards*, ed. Victor B. Brezik (Houston: Center for Thomistic Studies, University of St. Thomas, 1981). Joseph Owens, "The Christian Philosophy of *Aeterni Patris*," *Towards a Christian Philosophy* (Washington, D.C.: The Catholic University of America Press, 1990), pp. 63-75.

[4] McCool, *ibid.*, p. 196.

played in it. Gilson treated of the subject at length in *Recent Philosophy: Hegel to the Present.* In chapter 12, entitled "In the Spirit of Scholasticism,"[5] he discusses the origins of the neoscholastic movement, Leo XIII's call to revive Thomism, and the various flawed attempts to carry out the Pope's program. To neglect to consult this work of Gilson, as McCool does, is to miss one of the richest sources for Gilson's views on the subject.

In Gilson's words:

> The object of *Aeterni Patris* was to show that, indeed, the Church has never ceased to put natural reason at the service of Christian faith, either in order to defend it, or in order to elucidate its meaning. Before recalling the past history of Christian speculation, Leo XIII observed that such is, indeed, the best way of philosophizing, namely, to combine the religious obedience to faith with the exercise of philosophical reason. That is what was done by the early Christian apologists, the Fathers of the Church, and, finally the great masters of medieval scholasticism. The greatest among them was Thomas Aquinas. As a matter of fact, ever since the end of the thirteenth century, his authority has been recognized by popes and councils in many different solemn occasions. The Pope concluded that the safest way to restore philosophical and theological unity was to return to the traditionally approved doctrine of Saint Thomas.[6]

Here in a nutshell are the reasons why Gilson approved so highly of the Pope's encyclical. He knew that, like all encyclicals, *Aeterni Patris* "bears the mark of the time when it was written."[7] The Pope prescribed the

---

[5] E. Gilson, T. Langan, A. Maurer, *Recent Philosophy: Hegel to the Present* (New York: Random House, 1962), pp. 330-54.

[6] *Ibid.*, p. 339.

[7] *Loc. cit.*

teaching of the traditional philosophy of the Church, especially in its finest form, namely Thomism, but in the nineteeth century he could not be aware of the great diversity among the teachers of that philosophy. It should be noticed, however, that the Pope did not conceive the traditional doctrine as monolithic; he speaks of the variety of ways in which it has been handed down, for example by Bonaventure and Thomas Aquinas.[8] But the Pope could not be expected to take into account the results of the research into patristic and medieval philosophy that Gilson, among many others, carried out in the twentieth century, leading to the discovery of the striking originality and diversity of medieval philosophers and theologians. Indeed, it was in part owing to the Pope's injunction that Catholic historians of philosophy and theology undertook that research.

What first caught Gilson's attention in the encyclical and aroused his great admiration for it was its emphasis on Christian philosophy, not so much as a doctrine, as a special *way* of philosophizing. Certainly the Pope conceived of a traditional Christian philosophy in the Church, but he insists on the special mode that is appropriate to that philosophy. At the very beginning of his encyclical he announces, almost as its subject, that he wishes to speak about "the mode of taking up the study of philosophy which shall respond most fitly to the excellence of faith, and at the same time be consonant with the dignity of human science."[9] He describes that mode or method as uniting to the light of reason the superadded light of faith. The light of faith, far from

---

[8] *Aeterni Patris* (14), 42. I am citing the encyclical in the translation edited by E. Gilson, *The Church Speaks to the Modern World: The Social Teachings of Leo XIII* (Garden City, New York: Doubleday, Image Books, 1954). There is another translation in *One Hundred Years of Thomism*, pp. 173-97.

[9] *Aeterni Patris*, *ibid.*, (1), p. 32.

weakening or destroying human reason, completes and strengthens it and makes it capable of greater things.[10]

The reader of the encyclical can hardly fail to notice its repeated reference at the beginning to the mode, method, and right use of philosophy. Its right use leads the way to the true faith; it aids one to understand it better and to defend it against errors. This is said in no way to destroy the independence and integrity of philosophy for it enjoys its own method, principles and arguments.[11]

As McCool shows, Gilson's study of the medieval philosophers convinced him that they did not share a common doctrine, but they all philosophized in the same Christian way, uniting the light of faith to the light of reason, thereby gaining new philosophical insights. Unlike Descartes and modern rationalists, they did not separate their philosophies from religion but philosophized under its influence. Thus medieval philosophers were united not by a common doctrine but by the same spirit. This spirit is not the name of a philosophical system but of a special *way* of doing philosophy, which Gilson called Christian philosophy.[12] In his Gifford Lectures of 1931-1932, published under the title *The Spirit of Mediaeval Philosophy*, Gilson says: "Thus I call Christian, *every philosophy which, although keeping the two orders [of faith and reason] formally distinct, nevertheless considers the Christian revelation as an indispensable auxiliary to reason.*"[13] He contends, moreover, that the support revelation gives to philosophy in no way diminishes its rationality but rather enhances it. "A true philosophy," he writes, "taken

---

[10] *Ibid.* (2), p. 33.

[11] *Ibid.* (8), p. 37.

[12] G. McCool, *From Unity to Pluralism*, pp. 167-68.

[13] E. Gilson, *The Spirit of Mediaeval Philosophy*, trans. A.H.C. Downes (New York: Scribner's, 1936), p. 37. (Gilson's emphasis).

absolutely and in itself, owes all its truth to its rationality and to nothing other than its rationality."[14]

Reading the chapters on Christian philosophy in *The Spirit of Mediaeval Philosophy*, one would assume that Gilson was taking his inspiration from *Aeterni Patris*. He tells us, however, that when he wrote the book he had not yet read the encyclical.[15] What must have been his satisfaction to find his notion of Christian philosophy supported by an encyclical of the Church! As a Catholic, he regarded the document as the voice of the Catholic Church itself and not only as the expression of the personal views of Leo XIII. Encyclicals, he writes, "are the highest expression of the ordinary teaching of the Church."[16] It goes without saying, therefore, that for Catholic philosophers the importance of *Aeterni Patris* is much more than a merely historical one. Like all the encyclicals of the Pope, it should be read "as expressing, in its very essence, the thought of a twenty-centuries-old religious body."[17] In contrast, McCool looks upon the encyclical as "a purely disciplinary document, resting upon the juridical authority of the reigning pontiff, Leo XIII. Its scope was limited to the method of philosophical instruction

---

[14] *Ibid*., p. 40. In the preface to *Le Thomisme: Introduction à la philosophie de Saint Thomas d'Aquin*, 6th ed. (Paris: J. Vrin, 1965), p. 7, Gilson calls the philosophy of St. Thomas by its very nature "strictement rationelle."

[15] E. Gilson, *The Philosopher and Theology*, trans. C. Gilson (New York: Random House, 1962), p. 180. In an earlier work he says he did not remember the encyclical. See *Christianity and Philosophy*, trans. R. MacDonald (New York: Sheed & Ward, 1956), pp. 93-94.

[16] E. Gilson, *The Church Speaks to the Modern World*, p. 4.

[17] E. Gilson, *Recent Philosophy: Hegel to the Present*, p. 338.

approved for the education of future priests in seminary and Catholic faculties."[18]

McCool rightly remarks that, for Gilson, Christian philosophy is not a philosophical system but a way of philosophizing.[19] Gilson insists that it can be described but it cannot be defined as having a specific nature or essence. In this respect it is unlike philosophy or theology. These are two formally or essentially distinct wisdoms which can be defined in terms of their principles and methods. Christian philosophy does not add to them another wisdom with its own definable essence--a kind of hybrid that would combine and, in the process confuse, faith and reason.[20] Critics who accuse Gilson of creating such a notion of Christian philosophy fail to see that he does not locate it in the essential or formal order but in the order of mode or method. It is the way of philosophizing used by the Church Fathers and medieval schoolmen and productive of the brilliant philosophies of Augustine, Bonaventure, and Thomas Aquinas.[21]

Gilson found himself in agreement with *Aeterni Patris* not only on the notion of Christian philosophy but also on the superiority of Thomism over other Christian philosophies. The last third of the encyclical praises the philosophy of St. Thomas--without excluding others--as the model of the Christian way of philosophizing. The aim of the encyclical was the restoration in Catholic schools of Christian philosophy in the spirit of Thomas

---

[18] *Catholic Theology in the Nineteenth Century*, p. 1. For a more adequate description of *Aeterni Patris* see G. McCool, *Catholic Theology in the Nineteenth Century*, pp. 226-40.

[19] *From Unity to Pluralism*, pp. 167-68.

[20] E. Gilson, *The Philosopher and Theology*, pp. 182-83. *Christianity and Philosophy*, p. 95. "What is Christian Philosophy?," *A Gilson Reader*, ed. Anton C. Pegis (Garden City, New York: Doubleday, 1957), p. 187.

[21] E. Gilson, *The Philosopher and Theology*, pp. 175-99.

Aquinas.[22] The revival of Thomistic philosophy was intended to bring unity into the teaching of philosophy in Catholic schools, where the influence of Descartes, Leibniz, Kant, and other modern philosophers was rampant. A revised form of Aristotelianism was also being taught as a philosophy separated from the Catholic faith and theology. Leo XIII saw the revival of Thomism as necessary for unity in the teaching of Catholic philosophy and as a basis for reform in the Church's political, moral, and economic doctrines.[23]

But where is the philosophy of St. Thomas to be found? In his commentaries on Aristotle, in his theological writings or in both? Leo XIII insisted that it be drawn from the works of Aquinas himself without specifying which ones were to be used. Gilson contended that Thomas's most distinctive and personal views in philosophy are to be found, not in his commentaries on Aristotle, but in his theological works. I find no evidence for McCool's statement that Gilson thought he found this in *Aeterni Patris*.[24] He came to this conclusion from his own investigation of Thomas's philosophy.

Gilson's attitude toward Thomas's Aristotelian commentaries has often been questioned.[25] Did he

---

[22] In his apostolic letter of 1880 Leo XIII called *Aeterni Patris* "Our encyclical letter on the restoring in Christian schools of Christian philosophy according to the mind of the angelic doctor St. Thomas Aquinas" *(A.S.S.*, 13, 56). See E. Gilson, *The Church Speaks to the Modern World*, p. 6.

[23] E. Gilson, *The Church Speaks to the Modern World*, pp. 6-20.

[24] *From Unity to Pluralism*, p. 196.

[25] On St. Thomas as a commentator on Aristotle see J. Owens, "Aquinas as Aristotelian Commentator," in *St. Thomas Aquinas 1274-1974: Commemorative Studies* (Toronto: Pontifical Institute of Mediaeval Studies, 1974), 1, pp. 213-38. F. Van Steenberghen, *La philosophie au XIIIe siècle* (Louvain:

disqualify them as sources of Thomas's philosophy, as McCool says?[26] Statements of Gilson, taken out of context, would seem to support this. For example, he writes: "Saint Thomas is only a commentator in his writings on Aristotle. For his personal thinking one must look at the two *Summae* and similar writings, in which he shows himself an author in the proper sense of the word."[27] The context of this statement is the medieval distinction between a commentator and an author. According to Bonaventure a commentator "adds to a text only what is necessary to make it understood," whereas an author's "chief intention is to express his own thought...."[28] By and large this is a good description of Thomas's roles as a commentator and an author. Other statements of Gilson show that he esteemed the commentaries of Aquinas but that he regarded them as giving an imperfect and less personal account of his philosophical views than his theological works. The following lines from Gilson's *Le Thomisme* give his balanced opinion on the subject:

> The *Commentaries* of St. Thomas on Aristotle are very precious documents for us and their loss would have been deplorable. Nevertheless, if they had all perished, the two *Summae* would still preserve all that is most personal and most profound in his philosophy, whereas, if the theological works of St. Thomas had been lost, would we be as well informed about his philosophy by his commentaries on Aristotle?[29]

Publications Universitaires, 1966), pp. 310, 330, n. 33. J. Wippel, *Metaphysical Themes in Thomas Aquinas* (Washington, D.C.: The Catholic University of America, 1984), pp. 26-7.

[26] *From Unity to Pluralism*, p. 169.

[27] *The Philosopher and Theology*, p. 211.

[28] *Ibid.*, pp. 210-11.

[29] *The Christian Philosophy of St. Thomas Aquinas*, p. 8. The last sentence is slightly altered in *Le Thomisme*, 6th ed. 15.

Again,

> ...the commentaries on Aristotle, bound to follow the meanderings of an obscure text, only let us suspect imperfectly what might have been the nature of a *Summa* of Thomistic philosophy organised by St. Thomas himself, with all the sparkling genius that went into the *Summa Theologiae*.[30]

Gilson has also been taken to task for expounding Thomas's philosophy in the theological order, beginning with God and ending with creatures, instead of following the properly philosophical method of ascending from creatures to God.[31] McCool seems to be in sympathy with this widespread criticism.[32] But Gilson never gave in on this point. "In order to ascribe to Saint Thomas a philosophical mode of exposition," he writes, "I should have had to make it up."[33] As an historian he did not think his role was to invent but to understand. He had learned at the Sorbonne that an historian of philosophy should not create a doctrine and ascribe it to a philosopher but on the contrary attribute to him only what he himself thought and said. If it is true that Thomas created his philosophy as a theologian, and in the service of theology, it should be expounded in the theological order. Failing to do so is the reason so many "Thomists" have missed the deepest meaning of Thomas's doctrine. If a Thomist wishes to recreate that doctrine in a philosophical order he does so on his own responsibility and not on that of Thomas himself.

---

[30] *The Christian Philosophy of St. Thomas Aquinas*, p. 22.

[31] See F. Van Steenberghen, *La philosophie au XIIIe siècle*, pp. 346-54. James Collins, "Toward a Philosophically Ordered Thomism," *Three Paths in Philosophy* (Chicago: Regnery, 1962), pp. 280-99.

[32] *From Unity to Pluralism*, pp. 196-97.

[33] *The Philosopher and Theology*, p. 97.

Thomists often extract from Thomas's theological works sections that treat of philosophical matters and consider them as pure philosophy.[34] But in Gilson's view, Thomas's philosophy should not be separated from its theological context and order. Can the treatise on the human person (*De homine*) in the *Summa Theologiae* (I, 75-90) be regarded as a purely philosophical tract? In the prologue Thomas tells the reader that he is considering the person from the perspective of the theologian, with the consequence that he will treat primarily of the soul and only secondarily of the body insofar as it is related to the soul. The philosopher would do just the opposite. Thomas philosophizes deeply in this treatise but its organization is theological rather than philosophical. The same can be said of the treatise on law (I-II, 91-108), where Thomas takes up eternal law before natural and human law. A philosophical discussion of law would follow the opposite course. It is hazardous indeed to separate any parts of Thomas's *Summa Theologiae* and consider them as purely philosophical, for they all bear more or less explicitly the mark of their theological setting.[35]

Gilson has shown convincingly that Thomas's vocation was that of a theologian, not that of a philosopher. He philosophized abundantly and with great originality, but he did not think a Christian should stop at philosophy. To Thomas, philosophy was a help and consolation to the Christian on the road to salvation, and a vehicle for defending his faith and removing errors.[36] It afforded him valuable insights and truths

---

[34] See E. Gilson, *The Christian Philosophy of St. Thomas Aquinas*, pp. 21-22, 444, n. 58.

[35] See *The Philosopher and Theology*, pp. 94-95; *The Christian Philosophy of Saint Thomas*, p. 444, n. 58.

[36] See St. Thomas, *Contra Gentiles*, I, p. 9 (2). Arguments should be brought forth in order to make divine truth known. This should be done for the training and consolation (*solatium*) of the faithful. For the Christian's relation to

that are of service in the understanding of faith, which is theology. Theology itself is but a stepping stone to the final goal of human life, the loving vision of God.

Thomas' methodology in philosophy and theology is sometimes called "Aristotelian." McCool, for example, often speaks of Aquinas's "Aristotelian science of theology."[37] It is true that Thomas borrowed much from Aristotle, as he did from other ancient and medieval philosophers. But he transformed everything he borrowed for use in his theology. As Chenu and others have shown, Thomas devised his own notion of theology as a science.[38] Its proper name is not Aristotelian but Thomistic.

What did Gilson think his work as a Thomist should be? Very simple, according to McCool. "For the authentic disciple of the Angelic Doctor," he writes, "progress can consist only in his ever-deepening understanding of what St. Thomas himself has written."[39] Leo XIII had a broader vision in his program for restoring the wisdom of St. Thomas. The publication of new and better editions of his works and a more thorough understanding of his philosophy and theology were certainly at the heart of that program. But the Thomist was not only to look back to the past and undertake historical studies; the whole point to this

---

philosophy see Mark D. Jordan, "The Alleged Aristotelianism of Thomas Aquinas." (The Etienne Gilson Lecture 1990; Toronto: Pontifical Institute of Mediaeval Studies), pp. 38-39.

[37] *From Unity to Pluralism*, pp. 1, 2, 228. I find no evidence for McCool's statement in his *Catholic Philosophy in the Nineteenth Century*, p. 229: "Using Aristotelian science as its model, *Aeterni Patris* described philosophy as the organizing structure through which the various parts of theology were unified into a single interrelated body of knowledge."

[38] See M.- D. Chenu, *La théologie comme science au XIIIe siècle*, 3rd. ed. (Paris: J. Vrin, 1969), especially pp. 82-83. See also M. Jordan, *The Alleged Aristotelianism of Thomas Aquinas*.

[39] *From Unity to Pluralism*, p. 197.

endeavor was to look to the future and provide a more solid foundation for the social, political and economic structure of society.[40] The Pope placed *Aeterni Patris* at the head of his long list of encyclicals as a guide for restoring the social order in the light of Thomistic principles, under the aegis of the Catholic Church.

Moreover, the Pope did not conceive of the Thomist as turned in upon himself, as though other philosophers and scientists have nothing to teach him. Thomists were to open themselves to new ideas and discoveries for the enrichment of their philosophy. The Pope had "no intention of discountenancing the learned and able men who bring their industry and erudition, and, what is more, the wealth of new discoveries, to the service of philosophy."[41] Again, he writes: "We hold that every word of wisdom, every useful thing by whomsoever discovered or planned, ought to be received with a willing and grateful mind."[42]

Reading McCool, we would not think that Gilson had anything to do with this forward-looking program. In fact he heartily embraced it, contributing many monographs and articles to what he called "living Thomism."[43] He was a Thomist in the modern world, engaged in its religious, philosophical, moral, and cultural problems. In his voluminous bibliography there is a surprisingly large number of items devoted to these and other contemporary issues.[44] Of particular note in

---

[40] See E. Gilson, *The Church Speaks to the Modern World*, p. 5.

[41] *Aeterni Patris* (24), p. 47.

[42] *Ibid.* (31), p. 50.

[43] See E. Gilson, *The Spirit of Thomism* (New York: Kenedy & Sons, 1964), ch. 4 "A Living Thomism." See also the article of Victor Brezik in the present volume.

[44] See Margaret McGrath, *Etienne Gilson: a Bibliography* (Toronto: Pontifical Institute of Mediaeval Studies, 1982).

this connection are his three books on art,[45] his book on linguistics and philosophy,[46] and his book on evolution and finality in nature.[47] These are works of a Christian and Thomist philosopher, in which philosophy is not placed in the service of theology (as a professional theologian like Aquinas would do), though it remains open to the influence of Christian revelation and to the guidance of theology.[48] In these works Gilson philosophizes within faith but he uses the method and order of philosophy, not those of theology. Gilson insists that the theological order should be used when expounding the thought of St. Thomas, but I find no evidence for McCool's statement that, according to Gilson, Christian philosophy must always follow the theological order.[49]

Often overlooked is Gilson's quite novel method of philosophizing on the data of the history of philosophy--a method he used so successfully in books

---

45 *Painting and Reality* (New York: Pantheon Books, 1957); *The Arts of the Beautiful* (New York: Scribner's Sons, 1965); *Forms and Substances in the Arts*, trans. S. Attanasio (New York: Scribner's Sons, 1966).

46 *Linguistique et philosophie: Essai sur les constantes philosophiques du langage* (Paris: J. Vrin, 1969).

47 *From Aristotle to Darwin and Back Again: a Journey in Final Causality, Species, and Evolution*, trans. John Lyon (Notre Dame, Indiana: Notre Dame University, 1984).

48 Gilson's good friend and colleague, Anton Pegis, contends that the modern Thomist who a philosopher and not a professional theologian must distinguish between philosophy and its status as a handmaid to theology. He can be a Christian in his philosophy, opening his philosophy to the influence of revelation and to the guidance of theology, without being a theologian. See A. Pegis, *Christian Philosophy and Intellectual Freedom* (Milwaukee: Bruce, 1955), pp. 70-71. Gilson's works cited above, among others, answer this description.

49 *From Unity to Pluralism*, pp. 169-70.

like *The Unity of Philosophical Experience*[50] and *Being and Some Philosophers.*[51] These are often taken to be works in the history of philosophy, but in fact they are philosophical, using experience in the history of philosophy as a basis for philosophical reflection. From a sort of "experimentation" with that history Gilson draws intelligible ideas and laws that, in his estimation, transcend history and belong to philosophy itself.[52] Here too we find him writing as a Christian philosopher and not as a theologian.

This article does not address the wider issues of McCool's book. Like his previous volume, *Catholic Theology in the Nineteenth Century*, it is highly informative and very helpful for an understanding of the development in Catholic philosophy and theology in the nineteenth and twentieth centuries. The present article is limited to considering the pages in *From Unity to Pluralism* devoted to Gilson's notions of Christian philosophy and Thomism and their relation to *Aeterni Patris*. It is with regret that on these subjects I have found it necessary to question some of the author's judgments.

---

[50] (New York: Scribner's Sons, 1937).

[51] 2nd ed. (Toronto: Pontifical Institute of Mediaeval Studies, 1952).

[52] Gilson describes this method of philosophizing in "Remarques sur l'expérience en métaphysique," *Actes du XIe Congrès international de philosophie*, Brussels, August 20-26, 1953, vol. 4 (Amsterdam: North-Holland, 1953), 5-10. See the translation and study of this article by Armand Maurer, "Gilson's Use of History in Philosophy," *Thomistic Papers V*, ed. Thomas A. Russman (Houston: Center for Thomistic Studies, University of St. Thomas, 1990), pp. 25-48.

MARITAIN'S REALISTIC DEFENSE OF THE IMPORTANCE OF THE PHILOSOPHY OF NATURE TO METAPHYSICS

Raymond Dennehy

Introduction

In his book, *From Unity to Pluralism*, Father McCool argues that, contrary to the portrait of a single, unified Thomism painted by Leo XIII in *Aeterni Patris*, there are, in fact, several. As examples, he offers the differing interpretations of Thomas Aquinas' thought offered by Rousselot, Maréchal, Gilson and Maritain.[1] To support his claim that Maritain's brand of Thomism deviates from Aquinas', McCool argues that, among other things, Maritain was misled by Cajetan's three degrees of abstraction into supposing that metaphysics must be approached through the philosophy of nature rather than "...through a reflection on the intelligibility of mind."[2]

The counter-argument that is advanced in this essay is in two parts. The first is preliminary to the second and main part: too much has been made of the fact that, in his commentary on Boethius' *De Trinitate*, Aquinas uses the act of "separation" to indicate how the being of metaphysics is arrived at, while Cajetan denominates the act as "abstraction." The second part of the counter-argument is that an analysis of Maritain's thought fails to support McCool's claim that that

---

[1] Gerald A. McCool, S.J., *From Unity to Pluralism* (New York: Fordham University Press, 1992).

[2] *Ibid.*, p. 155.

"discrepancy" in terminology "cuts the ground out from under Maritain's argument that metaphysics must be approached through the philosophy of nature."[3] Indeed, Maritain's argument is more in line with Thomistic realism than its contrary, to wit, that the formal object of metaphysics, "being as being," is arrived at independently of the philosophy of nature. Whatever else Thomism may be, it is realistic. It draws its life from the principle, "The material object of the human intellect is the essence of sensible being"; put in other terms, "There is no knowledge in the intellect that does not come through the senses." Any representation of Thomism that collides with that principle is surely mistaken, for it is not realistic.

Accordingly, it is more plausible that Maritain's claim that the philosophy of nature is a prerequisite to the study of metaphysics springs from his understanding of Thomism *as a realism* rather than from any influence that Cajetan's three degrees of abstraction may have exerted on his thinking.

I

In From Unity to Pluralism, McCool writes:

> Cajetan's three degrees of abstraction provide the epistemological justification for Maritain's three ascending levels of speculative science. Maritain's analogy of science and its intelligibility rest on the three degrees of abstraction. His requirement that metaphysics be approached through the philosophy of nature requires Cajetan's three ascending degrees of formal intelligibility to justify it. The necessity of that approach to metaphysics is one of the main reasons for Maritain's total rejection of any attempt to

---

[3] *Ibid.*, p. 156.

> vindicate metaphysics through a reflection upon the intelligibility of mind.[4]

But "Unfortunately for Maritain," McCool continues, recent scholarship clearly shows that Cajetan's portrayal of the three degrees of abstraction fails to harmonize with the position of St. Thomas. The latter acknowledges abstraction on the levels of the philosophy of nature and mathematics, *abstractio totius* and *abstractio formae*, respectively:

> The being of the metaphysician is not grasped by abstraction at all. It is grasped through a negative judgment, the *separatio*, in which the mind affirms that all being is not material. Contrary to Cajetan's belief, analogy is not known prior to the metaphysician's grasp of being. Far from being a necessary condition for being's proper understanding, analogy itself is not understood until after the metaphysician has grasped being through the 'separation' of his negative judgment. Thus the new interpretation of St. Thomas' own thought cuts the ground out from under Maritain's argument that metaphysics must be approached through the philosophy of nature.[5]

In the first place, McCool makes too much of the difference between "separation" and "abstraction." It is not indefensible to urge, as does Maritain, that "separation" is but a different form of abstraction where "abstraction" is used analogously when applied to the three degrees of abstraction.[6] Whereas the abstractions

---

[4] *Ibid.*, p. 155.

[5] *Ibid.*, p. 156.

[6] Jacques Maritain, *Existence and the Existent*. Tr. by Lewis Galantiere and Gerald B. Phelan (New York: Pantheon Books, Inc., 1948), pp. 28-30, n. 14. See Armand Maurer's introduction to St. Thomas Aquinas, *The Division and Methods of the Sciences*. Tr.

employed in the philosophy of nature and mathematics are the products of simple apprehension, the abstraction used by metaphysics is the product of a negative judgment.[7] If abstraction indicates a focus of attention, why does not the negative judgment, "Being is not of its nature material," that yields the being of the metaphysician, qualify as an "abstraction"?[8] The distinction between "abstraction" and "separation" that Thomas Aquinas introduces in his *Commentary on the De Trinitate* of Boethius does not appear in his later works. In the *Summa Theologae*, for example, he writes of "two modes of abstraction," the one through judgment and the other through simple apprehension.[9] Even in the *Commentary on the De Trinitate* he uses the verb "to abstract" to indicate the act of "separating."[10] The rationale for Aquinas' distinction between "abstraction" and "separation" seems to be his growing recognition of the profound connection between the act

---

by Armand Maurer (Toronto: The Pontifical Institute of Mediaeval Studies, 1963), p. xxiv.

[7] See Maurer's use of Maritain's "eidetic visualization" in *Division and Methods of the Sciences*, p.xxiii.

[8] Maurer, *op. cit.*, p. xix.

[9] "...abstrahere contingit dupliciter. Uno modo, per modum compositionis et divisionis; sicut cum intelligimus aliquid non esse in alio, vel, esse separatum ab eo. Alio modo, per modum simplicis et absolutae considerationis; sicut cum intelligimus unum, nihil considerando de alio." Thomas Aquinas, *Summa Theologiae*, I, 85, *ad* 1; also see *ad* 2.

[10] "Et quia veritas intellectus est ex hoc quod conformatur rei, patet quod secundum hanc secundam operationem intellectus abstrahere non potest vere quod secundum rem coniunctum est, quia in abstrahendo significaretur esse separationem secundum ipsum esse rei sicut si abstraho hominem ab albedine, dicendo 'Homo non est albus,' significo separationem esse in re. Unde si secundum rem homo et albedo non sunt separato, erit intellectus falsus. Hac igitur operatione intellectus vere abstrahere non potest nisi ea quae sunt secundum rem separata; ut cum dicitur 'Homo non est asinus.'" Aquinas, *In de Trin.*, V, 3c.

of judgment and the act of existence in addition to the existential nature of the being of metaphysics.[11]

Nevertheless, for the sake of argument, let it be granted that being, as investigated by the metaphysician, does not come by way of abstraction. It is still very far from clear how much ground that actually "cuts out" from under Maritain's position that the philosophy of nature is a propaedeutic to metaphysics. Thomas Aquinas is crystal clear in enunciating, as he frequently does in his *Commentary on Boethius' De Trinitate* that all our knowledge comes from the apprehension of sensible being.[12] This has to include the metaphysician's knowledge of being. To say that the formal object of metaphysics, *being as being*, is attained by a "separation" rather than an abstraction simply tells us, as noted above, that the aspect under which the metaphysician investigates reality is arrived at from the realization that nothing in the nature of being requires that it be sensible or material. The material object of metaphysics, sensible being, without which the human intellect would never arrive at *being as being*, is derived from our knowledge of sensible things.

Aquinas' use of the *separatio* seems to flow from that doctrine. Our intellect directly conceives the essences of sensible reality but not of intelligible reality.[13] The latter, Aquinas writes, is arrived at by a process of negating sensible properties:

> Accordingly, we cannot say that we know immaterial substances obscurely by knowing their genus and observable accidents. Instead of knowing the genus of these substances, we know them by negations; for example, by understanding that they are immaterial, incorporeal, without shapes and so on. The more negations we know of them the less

---

[11] Maurer, *op. cit.*, pp. xxv-xxvi.

[12] *In de Trin.*, VI, 2.

[13] *In de Trin.*, VI, 3.

> vaguely we understand them, for subsequent negations limit and determine a previous negation as differences do a remote genus.[14]

If McCool's critique of Maritain's treatment of the degrees of abstraction has any significance at all for his claim that Thomism is a pluralism and not a unity, it is to show that Maritain has failed in his attempt at a "...total rejection of any attempt to vindicate metaphysics through a reflection upon the intelligibility of the mind."[15] But that failure, if it be such, cannot amount to much. For the vindication of metaphysics by reflecting on the intelligibility of the mind works within the context of a realistic metaphysics and epistemology only if based on a direct and certain knowledge of extramental being, which is to say, sensible being. Thus the plurality that McCool argues for amounts to very much less than the errancy that he imputes to Maritain's interpretation of the degrees of abstraction.

This is not to say that McCool's criticism of Maritain's grasp of how Aquinas differentiated the levels of human knowledge is misguided or lacking in importance. But what it does say is that, in the first place, a haze enshrouds the claim that there is a plurality of Thomisms: in what sense are we to understand "pluralism" here? Is it that Thomisms constitute different schools of philosophy in the way that Augustinianism, Thomism and Scotism are different schools, etc.? But then we could hardly get away with calling them all "Thomism." If we can apply the term seriously to the various "Thomisms," the reason must be that they possess enough important doctrines in common so as to constitute a unified intellectual view. They would then constitute a unity *per se* and a pluralism, *per accidens*. Although this might well rob "pluralism" of the significance with which McCool wishes to invest it, it

---

[14] Maurer, *op. cit.*, V, 3, p. 78.

[15] McCool, *op. cit.*, p. 155.

need not reduce it to unimportance. For example, making the starting point of philosophy anthropological rather than cosmological, that is, making the knowing subject rather than extramental being, the starting point, can be justified from the standpoint of realism in terms of contemporary, postCartesian sensibilities: a dialectical gambit designed to show that the operation of knowing dynamically strives toward being.[16] That is achieved by starting with the *cogito* as a self-validating proposition and then proceeding to show that the act of thinking is in itself incomplete and finds its completion in extramental being. But since it is impossible to proceed from thought to thing, it is clear that the only way to validate realism is to acknowledge the knowing subject's direct and certain knowledge of things. To use Maritain's formulation, "I know and what I know are things."[17]

In the second place, McCool's criticism of Maritain suggests a curious inadvertence to what it means to say that the material object of the human intellect is the essence of sensible being. That charter principle of

---

[16] See, e.g., Joseph Maréchal, *Le Point de départ de la métaphy*sique: Leçons sur le développement historique et théorique du problèm de la connaissance, 5 vols. (Paris: Desclee De brouwer, 1944-1949), Vol.V, pp.412 ff. and Louis De Raeymaeker, *The Philosophy of Being*. Tr. by Rev. Edmumd H. Ziegelmeyer, S.J. (St. Louis: B. Herder Book Co., 1966), pp. 23-28. But, as Henle decisively shows, when it comes to establishing extramental reality, even the Transcendental Thomists find it necessary to set aside their fascination with the Kantian *a priori* in favor of Aristotelian-Thomistic realism. Robert J. Henle, S.J., "Transcendental Thomism, A Critical Assessment," *One Hundred Years of Thomism*, edited by Victor B. Brezik, C.S.B. (Houston: Center For Thomistic Studies, 1981), pp. 90-116; see Raymond Dennehy, "The Ontological Basis of Certitude," *The Thomist,* 50 (1986), pp. 120-150.

[17] Jacques Maritain, *The Degrees of Knowledge*. Tr. by Gerald B. Phelan (New York: Charles Scribner's Sons, 1959), p. 76.

realism is doubtless what inspires Maritain's position on the importance of the philosophy of nature for metaphysics: although the former's focus on being in its corruptible and transient state relegates it to an inferior and subordinate position in relation to the latter,[18] its rejection as a necessary initial stage of philosophical inquiry before arriving at the investigation of *being as being* destroys the rational credibility of metaphysics, and it does so in three ways.

II

First, it severs the connection between the being of metaphysics from sensible reality. Simply to say that Thomas Aquinas arrives at the formal object of metaphysics by "separation" rather than abstraction does not invalidate Maritain's claims for the necessity of the philosophy of nature. For it is through the latter that being and its principles are educed and receive their "scientific," i.e., their philosophical, status. Because the starting point of all our knowledge is sensible being, the philosophy of nature is a crucial step in the *via inventionis*. To embark on the study of metaphysics by merely separating *being as such* from the concept of being produces no more than *logical being*, which, if taken as the being of the metaphysician, reduces metaphysics to vacuity. It would then be the being of the idealist philosopher, possessing only the properties of extensiveness and inner-coherence.[19] A metaphysics based on that conception of being has no anchor in the real world. The most that can be claimed for it is that it is an object of the philosopher's thought from which no valid inference can be drawn about real being. Is this not

---

[18] Maritain, *Science and Wisdom*. Tr. by Bernard Wall (London: G. Bles, 1940), p. 35.

[19] Maritain, *A Preface to Metaphysics* (New York: Books For Libraries, 1979), pp. 36-38.

the empty logical shell of the mind (*sinlos*) that Kant accused the metaphysician of mistaking for reality and which the empiricists suppose to be *being-in-itself*?

In his journey toward what he called "The True Subject of Metaphysics,"[20] Maritain exposed the "Counterfeit Metaphysical Coin,"[21] under which rubric he included the being of the logician or "Being Divested of Reality"[22] and "Pseudo-Being."[23] Being, for the logician is an *ens rationis* which, although presupposing the being of reality, i.e., being gained through a knowledge of the world, is nevertheless in itself a second intention produced by the mind's reflection on its own reasoning activities. As such, it is the being produced by *abstractio totalis*, a conception possessing the widest possible extension or universality.[24] In terms of its intention or comprehensible constituents, it is practically devoid of meaning.

If taken as the standard of being, this notion of being becomes "pseudo-being." The sacrifice of a concept's comprehension in favor of its extension, which is to "...neglect the characters which intrinsically constitute an object of thought,"[25] demands that being be construed purely in terms of its lesser or greater extension so that it is no different "from conceptual objects of a purely generic nature."[26] Being thereby loses its status as a transcendental and becomes a genus. Because all things are reducible to being, the concept of being is the broadest, most universal of all concepts; and because being is now construed as a genus, it becomes the emptiest of all concepts. The reason for this

---

[20] *Ibid.*, pp. 43 ff.
[21] *Ibid.*, pp. 17 ff.
[22] *Ibid.*, p. 33.
[23] *Ibid.*, p. 36.
[24] *Ibid.*, p. 35.
[25] *Ibid.*, p. 36.
[26] *Loc. cit.*

evisceration is clear from the manner of arriving at a genus: one eliminates from the definition of a genus all characteristic notes of species; from *animal*, for example, one refuses to introduce those notes of any one of its species, *man*, say. Now, if being is the supreme genus, it follows that one must eliminate *all* characteristic notes. The final product is a concept that possesses the widest possible extension and the least comprehension. Supposing this product to be the correct conception of being, Hegel found it impossible to distinguish being from nothing and therefore identified it with nonentity.[27] Similarly, Collingwood concluded that there was "no science of pure being" for the simple reason that such a "science" "would have a subject-matter entirely devoid of peculiarities; a subject matter, therefore, containing nothing to differentiate it from anything else, or from nothing at all."[28]

Because of its purely generic character, such a notion of being must not only be conceptually empty but univocal and static as well. As noted above, these make it indistinguishable from the being of the logician. It is a conception of being that obeys the laws of thought rather than of reality and which consequently severs all ties with the being of things. Thus independent of the laws of real things, it no longer presupposes the latter but is instead a pure form of thought.[29] Regardless of realistic commitments and personal intentions, the fact remains that this conception is of pseudo-being, because by its very univocity and static nature it claims no connection with real being. It belongs instead to the ultimate conclusion of rationalism in which, as we see in Parmenides, the diversity and dynamism of being are

---

[27] *Ibid.*, p. 37.

[28] R.G. Collingwood, *An Essay on Metaphysics* (Oxford: The Clarendon Press, 1957), p. 14.

[29] *Preface to Metaphysics*, p. 37.

sacrificed on the altar of intellectual unity and inner-coherence.

It is quite a different matter to arrive at the being of the metaphysician by *separation* after discovering being in sensible things and grounding the resulting notion of being and its principles in the only reality of which we have direct experience, sensible being. From the understanding of being as *"that which is or exists,"* it follows that materiality is not of the essence of being. This step can catalyze a fuller understanding of the connection between *being* and *immateriality* by leading to the realization that not only is being not in itself material but that the higher on the scale of being a things is, the less materiality plays a part in it. From the premise that being is "that which is," it follows that a thing has unity to the extent that it has being, for if being is that which is, then it is what it is and not another thing. The more being a thing has, the more it is what it is, the more, then, it possesses unity and thus the closer its approach to genuine uniqueness. Individuals are more generic and anonymous depending on how low the rung they occupy on the ladder of being. The reason for this is that matter is without a center; it is, as Aristotle observed, "parts outside of parts." Lacking a center, matter is extension; it occupies space. Hence, two rocks, say, lay claim to being individuals because, by the fact of their extensiveness, two bodies cannot occupy the same space at the same time. They are thus individuated not so much by virtue of their intrinsic properties but rather insofar as their respective extensions separate them from each other. Because matter is a principle of potency as well as individuation, the conclusion must be drawn that, to the extent that a thing is enmattered, to that extent is lacks being and thus unity; for just to that extent it is not being but has only the potency to be.

This explains why the higher the rung a thing occupies on the ladder of being, the more closely it approximates genuine uniqueness. Although individuals, rocks and plants do not have personalities, nor do animals; though in the latter case we do apply that term to them, but only after a manner of speaking. It is only when we ascend to the rung of human beings that genuine personality is encountered. That is so because matter does not dominate them. To be a person is to be a self, a unique center of conscious, autonomous being. And to be a self is, in turn, to be self-identical, a being who can utter to him or herself the word "I." This utterance implies the utterance, "I am I," because to apply that personal pronoun to oneself is to acknowledge one's perfect unity. It is a perfect unity in the sense that to know oneself as an "I" is to affirm one's self-identity. Such unity or reflexivity is possible only in the world of spirit. The feat of bending back on oneself, of performing an act of perfect reflexivity, cannot be accomplished by a material faculty since matter occupies space; two bodies cannot occupy the same space at the same time. Thus the uniqueness of the person is the unity of being; and the more unity a thing possess, the freer it is from matter. Being is that which is; that which is is what it is to the extent that it is; it is what it is to the extent that it has unity; and it has unity to the extent that it is being; but it has being to the extent that it is free from matter since matter is not being but the *potency to be*.[30]

In some such fashion, one could indeed arrive at an ever fuller understanding that the being of the metaphysician is not material being. But it is not likely that one would thus proceed simply on the basis of "...a reflection on the immateriality of mind." From Parmenides down to the present, the discounting of sense knowledge in favor of the unifying proclivities of the intellect has hardly led to an appreciation of the uniqueness of being. A univocal rather than an

---

[30] See Raymond Dennehy, "Being is Better Than Metaphysics," *Freedom in the Modern World.* Ed. by Michael D. Torre (Notre Dame: University of Notre Dame Press, 1989), pp. 256 ff.

analogous conception of being has been the preferred choice of rationalism.

In contrast, the journey to "The True Subject of Metaphysics" takes quite a different path. It is the path of the intuition of being as such. It is an abstractive intuition which Maritain prefers to call an "*eidetic* or *ideating visualization*" or "ideating intuition," by which he means "an intuition producing and idea..."[31] "Eidetic visualization" is thus an abstraction whereby

> ...the intellect by the very fact that it is spiritual proportions its objects to itself, by elevating them within itself to diverse degrees, increasingly pure, of spirituality and immateriality. It is within itself that it attains reality, stripped of its real existence outside the mind and disclosing, uttering in the mind a content, an interior, an intelligible sound or voice, which can possess only in the mind the conditions of its existence one and universal, an existence of intelligibility in act.[32]

In other words, the abstraction consists in focusing on the existent reality that all things have in common -- they exist. But this is an analogous, not a univocal notion of existence. It is, after all, derived from our existence of extramental being and the latter consists of things that are diverse in spite of their oneness or unity. As Maritain emphasized elsewhere, existence is not a generic quality, an injection of which into an essence causes it to cross over from nothingness to existence. On the contrary, existence is an act that is exercised; to exist is to exercise an act.[33] Thus, if we discriminate between sources of activity, i.e., subjects, and recipients of activity, i.e. , objects, then, from the

---

[31] *Preface to Metaphysics*, p. 61.

[32] *Loc. cit.*

[33] *Degrees of Knowledge*, p. 436; *Existence and the Existent*, p. 62.

metaphysical standpoint, only subjects populate the universe since existence is an act that is exercised.[34] The conception of a generic act is the grasp of the *essence* of act, *actness*, not the acts that existent subjects exercise; *a fortiori*, the conception of a generic act of existing constitutes the transformation of existence into an essence. That is why the logician's notion of being, if taken as the meaning of being itself, becomes pseudo-being: being is that which exists and that existence is not *actness* but *an act that is exercised.* But as noted above, a thing is unique to the extent that its existence dominates its essence; thus, things are at once many and one:"Being" presents me with an infinite intelligible variety which is the diversification of something which I can nevertheless call by one and the same name, because it is in all cases made known to me by the similar relationship which the most diverse objects possess to a certain term essentially diverse, designated in each by our concept of being, as being present formally and intrinsically in it. And this analogical character, an example of what is called the analogy of strict proportionality, is inscribed in the very nature of the concept of being.

> It is analogous from the outset, not a univocal concept afterward employed analogously. It is essentially analogous, polyvalent. In itself it is but a simple unity of proportionality, that is, it is purely and simply manifold and one in a particular respect.[35]

Maritain's concern to get across to his audience the significance of the analogical nature of being discloses itself in his translation of *abstractio totalis* and *abstractio formae* into the respective terms, "extensive visualization" and "intensive visualization." Initially, intellectual visualization is simply extensive. Instead of

---

[34] *Existence and the Existent*, p. 62.

[35] *Preface to Metaphysics*, p. 67.

explicitly grasping the essence of the thing as its object of knowledge, what the intellect grasps is a concept that is more or less general, with the essence only implicitly present; nevertheless, this nascent intellectualizing introduces the intellect to the intelligible order insofar as it now confronts the universal in general.[36]

Intensive visualization is the next step, by which the intellect grasps and lays bare the "universal type and essential intelligibility," and thereby separates the essence, that which formally constitutes the thing, from its contingent and material properties. Thus to know that the individual proceeding towards me is a human being is to apprehend in it the essence *man*, an intelligible object that is independent of his or her height, skin color, degree of intelligence or health, etc. This intensive or typifying visualization Maritain describes as "the beginning of scientific knowledge, *knowledge* in the strict sense."[37]

Intensive or typifying visualization is the degree of abstraction by which the sciences differ from each other. But this means that the object of a higher degree of intensive visualization is more universal at the same time as it is of a higher order.[38] For what makes the being of the metaphysician more universal is the very thing that accounts for its more perfected degree of intelligibility -- its greater ontological density. "Being as being" enjoys the widest possible universality of any concept not because of its extensivity but because of its intensivity. If the former, then, as noted earlier, it would belong to the province of logic and dialectic, not metaphysics; for, as it became more extensive, owing to successive abstractions, it would become ontologically emptier. On the contrary, "being as being" is the most universal of all concepts, not by virtue of what it leaves

---

[36] *Ibid.*, p. 77.

[37] *Loc. cit.*

[38] *Loc. cit.*

out but by virtue of what it leaves in. Being is not only that which is; it is all there is; outside being there is nothing. All things are thus reducible to being.

All of which is to say that what makes being analogical is exactly what makes it transcendental. The first grasp of being comes through the senses, but once the concept is explicitly entertained, even in its imperfect, material state, the intellect is in a position to progressively purify it until the recognition occurs that being is in itself absolutely perfect and infinite. For if being is all there is and what is outside being is nothing, then it follows that being itself transcends all limitations since it cannot be contained in a genus or category: nonbeing cannot limit or contain anything.

Because the ever wider universality revealed by the degrees of intensive visualization does not result from an extensive abstraction, the degrees of knowledge are characterized by an ontological continuity. Being comes to be known, in the first place, through the intellectualizing of sensorial data;. one never has a direct knowledge of any other beings than those grasped in the sensible world through the senses: corruptible, changing being. That is the ambit that gives birth to the knowledge of "being as being." In all their diversity, the things of the world show that they have being in common, but they have it in different ways.

Again, it all comes down to the question of how much bearing the principle, "No knowledge is in the intellect that does not come through the senses," has on the metaphysician's concept of being. The following points are accordingly worth considering:

1. The subject of the philosophy of nature is ontological since its focus is on being *insofar as it undergoes change*: In his commentary on Aristotle' *Physics*, Thomas Aquinas puts matters thus: "...its subject is mobile being simply. I do not say *mobile body*, because the fact that every mobile being is a body is proven in this book and no science proves its own

subject."[39] Why that particular focus? Because that is what we perceive in the world. The philosophy of nature came into existence as a response to the problems of the one and the many. Perplexed by the phenomenon of change, the early Greek philosophers asked, "If things assume different appearances, which appearance is the reality? Absent an analogous conception of being, the respective positions of Parmenides and Heraclitus express the inevitably polarized answers to the question. Parmenides' position follows from a univocal conception of being, while Heraclitus' follows from an equivocal conception. The empiricist tradition, insofar as it reduces all knowledge to perception and thus materializes the mind, would clearly embrace the pluralism of being seen as equivocal because of its nominalistic commitments. But a conception of being *as not material* arrived at from an initial conception of being whose rationally justifiable origins reach to the murkiness of "a reflection on the intelligibility of mind" will, for reasons already unfolded, inevitably find itself partners with the rationalist tradition, which would embrace the unity of being seen as univocal.

2. "Being as being" exists nowhere but in the mind of the metaphysician. In Maritain's terms, it is an "*eidetic visualization*." From the intuition of being, the realization that things *exist*, an analogous concept is formed which expresses both what real things have in common, that they are, they exist, and what differentiates them:"their essences and individuating factors: *a*'s existence is to *a* as *b's* existence is to *b*.[40] What exists outside the mind are particular beings, *"things, existents, subjects*."[41] "Being as being"

---

[39] Thomas Aquinas, *Commentary on Aristotle's Physics*. Tr. by Richard J. Blackwell, Richard J. Spath and W. Edmund Thirlkel (New Haven: Yale University Press, 1963), Lecture I (184 a 9 - b 14), p. 4.

[40] *Existence and the Existent*, p. 30.

[41] *Degrees of Knowledge*, p. 433.

expresses the analogous nature of being, a conception whose derivability from "a reflection on the intelligibility of mind" is precarious. The beings that we perceive in the world are diverse and dynamic rather than uniform and static as is a conception of being simply excogitated. The doctrine of the analogy of being originates in the observation that things have being and yet are diverse; they change and yet are permanent. In short, things display both unity and plurality.

"Being as being" gives us a knowledge of reality as such. We thus know things because knowledge follows from being. But the existential reality of being can only be grasped in actually existent things. A conception of being derived from the mind does not suffice. In the philosophy of nature, the mind wrestles with change and corruptibility -- it does not investigate them for their own sakes -- in order to study being, the real. Regrettable though it may be, the only beings we have at hand are changeable and corruptible.

III

The second way, according to Maritain, in which the rejection of the philosophy of nature as a preliminary to the investigation of "being as being" destroys the intellectual credibility of metaphysics is as follows. If one does not arrive at being through the philosophy of nature, then how can metaphysics justify its position as *scientia rectrix*? Surely it cannot judge the philosophy of nature and oversee the scope and limits of its inquiry if it has no ontological and thus no epistemological connection with it. For then it would be powerless to direct the philosophy of nature and the science of phenomena towards a knowledge of true wisdom since it would lack "an intellectual grasp of the real as such."[42]

---

[42] *Science and Wisdom*, p. 50.

A major consequence of such impotence would be, and is, the emergence of logical positivism with its commitment to science as the criterion of all our knowledge of the real world. Without the ontological and epistemological connections between metaphysics and the philosophy of nature, it is impossible to furnish a rationale for curbing this deification of scientific knowledge.

IV

The third way that Maritain sees metaphysics harmed by the rejection of the philosophy of nature as its propaedeutic is that it undermines metaphysics as the "...speculative knowledge of the highest mysteries of Being naturally accessible to our reason."[43] If the philosophy of nature presupposes metaphysics, *in the order of reality*, metaphysics presupposes the philosophy of nature *in the order of discovery*. Since all our knowledge comes to us through the senses, the philosophy of nature is the material basis for metaphysics. For, as noted above, it is the discipline in which we discover and rationally confirm being and its principles. Without it, metaphysics can claim no rationally defensible tie with things.[44]

Granted, "...we have immediate contact with the real only through our senses,"[45] but how does that make the philosophy of nature requisite to metaphysics? Is it not sufficient that "being as such," the common being investigated by metaphysics, be arrived at by means of a separation from the notion of sensible being? Why cannot the latter serve as the "material basis" for metaphysics?

---

43 *Ibid.*, pp. 48-49.

44 *Ibid.*, p. 49.

45 *Loc. cit.*

Maritain's answer seems to be that a *science* of being as such requires a science of being as sensible, which is to say, a science of being *insofar as the latter undergoes change*. But why must that be so? His answer is in two parts. First, given that our senses put us in contact with the real, then "...a knowledge of the pure intelligible, a knowledge situated at the highest degree of natural spirituality, cannot reach the universe of immaterial realities if it does not grasp first of all the universe of material realities."[46] That Maritain intends this imperative in a scientific, i.e., a philosophical, sense rather than in a merely experiential sense is clear from the second part of his answer:

> And it [a knowledge of the pure intelligible] cannot grasp the universe and unearth its proper object if a knowledge of the intelligible mingled with or overshadowed by the sensible is held to be impossible: by this I mean a knowledge inferior in spirituality which first of all attains the being of things in so far as it is clothed in mutability and corruptibility, and which thus prepares, announces and prefigures metaphysical truth in the shadows of this first degree of philosophical knowledge.[47]

Maritain seems to be arguing in the above that metaphysics requires an accurate, true-to-life notion of being and that the latter can be attained only by means of the purification of being, which, in turn, can come only through a knowledge of sensible being. That is, the being apprehended through the senses is not only being that is sensible but being that is changing. Thus to steer between the respective errors of Parmenides and Heraclitus, a philosophy of nature is necessary.

Aristotle successfully preserved both unity and diversity, change and permanence, by his hylemorphic

---

[46] *Loc. cit.*

[47] *Loc. cit.*

theory. Their reconciliation was made possible by furnishing a rational justification for what our experiential knowledge of the world reveals to us: the things that we experience display both sameness (unity) and difference (plurality), permanence and change.

Without the scientific validation of being as changing and diverse, what would the being of metaphysics be like? Its formal object, being as being, would lack the ontological richness that comes with diversity of being. It would instead be a vacuous concept insofar as the expression, "being as being," would then mean no more than an empty logical tautology, reducible to the merely logical formulation of the principle of identity, "A is A." Vacuity is price of robbing being of its analogical nature.

If there is no scientific (philosophical) justification for the analogy of being, then there is none for the diversity and changeability of being. The alternative is either monism or chaotic flux. As already discussed, the former if our experiential knowledge of the world is denied in favor of a *mere* concept of being distilled from the *mere* concept of sensible being; the latter, if being as unity is denied in favor of the diversity and dynamism of the sensible world reported by our perceptions.

The point is not that our experiential knowledge shows beyond a doubt ("It is certain and evident that some things in the world are in motion...", etc.) that the analogical nature of being is grounded in the nature of sensible things. It is rather that the mere appeal to our prephilosophical experiences of the world does not add up to a rational and philosophical justifiation for either the analogical nature of being or the grounding of the metaphysical notion of being in actual beings. Without rational justification for that grounding, we are left with the embarrassing Cartesian dichotomy between philosophy and commonsense.

What the absence of grounding does add up to is the replay of the polarization of philosophy into rationalism and empiricism; between genuflection before the canons of intellectual unity at the cost of sensible content and genuflection before the canons of sensible content at the cost of intellectual unity. That is the stage setting upon which Immanuel Kant entered. The polarization of concept and sensible data creates the crisis of the *a priori*: how reconcile the universal and necessary with the particular and contingent? The very affirmation of concepts not derived from sensations reduces the universal and necessary to the status of empty logical shells (*sinlos*) whose only claim to legitimacy is to be filled with sensible data.

If the metaphysician's concept of being does not derive from the changing, corruptible being of the philosophy of nature but is instead the rational distillate produced by its separation from the concept of material being (i.e., from the negative judgment, "Being in itself is not material"), one can appropriately ask, "Whence the concept of being that gives rise to it?" "Is it not a mere, empty logical shell characteristic of the *a priori* family?"

Conclusion

All in all, the plausible explanation of why Maritain insists that you cannot arrive at an authentic metaphysics except through the gate of the philosophy of nature is to be found in his profound appreciation of the principle, "The material object of the human intellect is the essence of sensible being," and not in any putative misinterpretations by Cajetan and the other commentators of Aquinas' basis for distinguishing the levels of knowledge. This appreciation could well have been inspired by his early master, Henri Bergson, who possessed such a firm grasp of the difference between the real and the conceptual, by his wife, Raissia, with her poet's sense of the intuition of the real, and by his lifelong struggle to meet the challenge, bequeathed him by Bergson, of reconciling conceptual knowledge and reality. From Maritain's claim that if one had to choose between the study of the sciences and the humanities, one should choose the sciences in order to keep the mind in touch with nature, to his preference for the company of missionaries, mystics and artists rather than professors and students,[48] to his sketch of the two-dimensional academic whose knowledge comes almost entirely from books, to his controversial claim that nobody, no matter "how freighted with erudition," can be a metaphysician without the intuition of being,[49] this appreciation of the real has pervaded the range of Maritain's life and interests. A study of the meaning and implications of realism in his writings would doubtless prove a fruitful contribution to Maritain scholarship. At all events, Maritain's unyielding position on the indispensability of the philosophy of nature for metaphysics must be evaluated within the context of the meaning and implications of the principle, "The material object of the intellect is the essence of sensible being."

---

[48] Yves R. Simon, "Jacques Maritain: The Growth of a Christian Philosopher," *Jacques Maritain: The Man and His Achievement,* ed. by Joseph W. Evans (New York: Sheed and Ward, 1963), p. 18.

[49] *Existence and the Existent*, p. 22; also p. 19.

APROPOS OF *FROM UNITY TO PLURALISM* BY GERALD MCCOOL, S.J.[1]

Robert J. Henle, S.J.

This is a brilliant book. Father McCool's comprehensive scholarship, his control of a vast amount of material, the clarity of his analysis and exposition are all extraordinary. No matter what criticism may be made of this book, it remains one not only well worth reading but also well worth studying. The book is so rich in subject matter and reference that it suggests a wide variety of studies. I shall emphasize, within a broader context, some metaphysical and epistemological concerns suggested by various parts of the presentation.

I. The Neo-Thomistic Program.

But now let us examine some of the major points of his thesis, especially as they relate to Thomistic metaphysics and epistemology.

Father McCool derives from *Aeterni Patris* and the expectations of the early promoters of the "Thomistic Movement" a rigid and highly ideological program for the "Movement."[2] This sort of Thomistic program is and should be gone. But McCool uses it rigidly as a basis for describing and evaluating the progress of Thomistic theology and philosophy since Leo XIII. The rather demanding use he makes of this original model leads to an obscuring of the unity of the Thomistic development and an initial devaluation of Thomistic theology and philosophy.

---

[1] Gerald A. McCool, *From Unity to Pluralism* (New York: The Fordham University Press, 1989).

[2] *Op.cit.*, pp. 1-5 *passim*; p. 23.

The original model assumed that there was a common and unified doctrine, "wisdom," among the great Scholastics. However, Gilson's[3] studies as well as the investigations of other scholars have shown that this assumption is false. Scholasticism displays a wide diversity of opinions and "systems." This McCool counts as one of the reasons for the decline of the movement. But from the standpoint of Thomism itself, these studies clarified and emphasized the unique qualities of St. Thomas' thought and his superiority among the Scholastic Doctors. We not only understand the insights of St. Thomas better but we can intelligently determine what valuable insights other Scholastics have to add to the Thomistic tradition itself.[4]

Again, the first promoters of the movement believed that the sixteenth and seventeenth century commentators and the leaders of the Thomistic revival in Spain and Portugal developed a presentation of authentic Thomism. This is now known to be questionable. According to McCool, here is another blow to the "Movement." But it has enabled us again to understand St. Thomas better and to advance a more enriched Thomism without introducing destructive errors into it. For example, the work of John Deely in elucidating the positions of John of St. Thomas (John Poinsot) is developing a new understanding of the potential of St. Thomas' philosophy of knowledge for current application of Thomism in modern philosophical movements.[5]

The original platform is said to have assumed that St. Thomas had distilled the wisdom of the Fathers and have presented that distillation in precise and accurate

---

[3] McCool lists Gilson's major studies on p. 257.

[4] On the unique character of St. Thomas' existentialism see McCool, *op.cit.*, pp. 182-186.

[5] John Poinsot, *Tractatus de Signis*, ed. and tr. by John Deely (Berkeley: University of California Press, 1985).

terms.[6] Numerous studies have shown that this is not so. And there is good reason why it must be so. This lies in the difference between the humanistic mode of knowing and communicating and the scientific mode. In the humanistic mode all the cognitive and affective powers of human nature are used to convey an experiential understanding. It concentrates on the singular, the concrete, the direct interpretation of lived human experience.[7] Now a great deal of Scripture and of the writings of the Fathers is in this mode. And it is never possible to translate humanistic understanding simply into philosophical understanding. Besides, studies like those of Henri de Lubac have shown the inadequacy of the assumption that the wisdom of the Fathers is to be found intact in St. Thomas. It is essential that there be continuous study of the Scriptures and of the Patristic writers not only to supply doctrinal material for theology but also to promote the religious life of the people. Thomistic theology needs this exploitation of these sources. The elimination of the idea that the study of St. Thomas or any other systematic theologian can substitute for study of the sources is all to the good of Thomistic theology.

The upshot of these considerations is that the disappearance of the original rigid protocol has not destroyed Thomism but rather has freed it from a false perception of it.

From now on I will deal mainly with philosophy with special emphasis on epistemology.

II. System or Synthesis?

Before proceeding, however, an important distinction must be made. Throughout the book McCool

---

[6] McCool, *op. cit.*, pp. 207-208.

[7] See R. J. Henle, S.J., *Theory of Knowledge* (Chicago: Loyola University Press, 1983), pp. 337-360.

seems to use the terms "system" and "synthesis" as essentially synonymous. I think they should be given a clearer meaning that will differentiate them.

A philosophical or scientific system is a body of knowledge organized in a deductive series of steps following from some first principles, postulates or facts. Systems tend to be highly abstract and univocal. A perfect example of a system is Euclidean geometry. A set of definitions and axioms (principles) is laid down. This set controls the subsequent deductions so that all subsequent theorems depend on the initial principles. And all are tied together by the necessity intrinsic to deductive reasoning. Modify one of the axioms and the entire system is changed.

Likewise, no new formality can be introduced within the chain of argumentation; that, too, simply destroys the system. In philosophy a good example of a system is Spinoza's *Ethica*, developed precisely in imitation of geometry. That system stands or falls as a whole and, as a whole, it is unbelievable, and therefore there are flaws in its principles.

Now many early exponents of the "Movement" presented Thomism as a system. This is reflected especially in the Latin manuals and their early vernacular counterparts. Thus condensed into a didactic manual, Thomism appeared to be abstract, rigid and *a priori*, with minimal support from induction or experience. This presentation also closed the system and made it unable to adapt to new problems or to adopt new ideas. In my opinion, it made Thomism a caricature of itself.

Therefore, I turn to the concept of a "synthesis." A synthesis is the putting together of different data, ideas and principles in various relationships to one another. There is no longer a central line of deduction. There may be pervasive principles or grand views that give some homogeneity to the synthesis. Thus in the Thomistic synthesis the doctrine of act and potency is pervasive, serving to illumine many different aspects of Thomism.

A synthesis is flexible; its parts can be recognized in various ways. Now ideas or data can be introduced into it without destroying it. The parts of a synthesis can have their own inductive or experiential source.

A miniature sample of the difference between "system" and "synthesis" may be found in two descriptions of conversation. The Neo-Scholastics did a systematic analysis of the movement towards faith:

> The key to the solution of this set of problems, Rousselot believed, lies in the metaphysics of knowledge. Neo-Scholastic theologians divided the movement toward faith into a succession of separate acts. There are a judgment of credibility, affirming the reasonableness of the act of faith, and a judgment of "credentity" affirming the moral obligation to make it. There is the will's free disposition to make the act of faith. Finally there is the supernatural act of faith itself.[8]

We have the rigid systematic approach, with little flexibility and little reference to the way conversations actually take place.

Rousselot's subsequent discussion shows that he is thinking of a "process" and of a "synthesis."

> Far from being a *succession of discrete acts*, the movement to faith is an *uninterrupted process*. It is an "*aperceptive* [sic] *synthesis*" into which these various elements enter as moments. The process of faith is not a *discursive* process of the *ratio*. It is a movement of the *intellectus*, the higher power of insight. The influence of connaturality and the *attitude* of sympathetic love in the "aperceptive [sic] *synthesis*" can account for the reasonableness and freedom of the supernatural act of faith.[9]

---

[8] McCool, *op. cit.*, p. 76.

[9] *Ibid.*, pp. 76-77.

Thus there is a varying interaction between all the elements of human knowing, willing, and loving, combining into various individual syntheses.

> It would pay the theologian to observe how scientific induction works. In his observed facts the scientist intuitively grasps the *clue* to his general law. In a sense the clue is prior to the law since it leads to it; yet, in another sense, the law is prior to the clue. For it is only in the light of the general law that the clue can be seen as a clue and that its significance in the law's intelligible pattern can be recognized. *Intellectus* is not discursive *ratio*. It grasps whole intelligible patterns in a moment of insight. In that moment fact and law illuminate each other reciprocally. Facts can be observed but they do not become clues until the instant in which the law is grasped.[10]

A synthesis of this sort does not finalize a position, cutting off any additional clues or insights. Note that for faith a clue can be anything. Rousselot says it may be an "encounter with a saintly priest."[11] It could be the loving of a Catholic spouse, a thoughtful book, etc.

Because Rousselot's account has the flexibility and openness of a synthesis and not the rigidity of a system, it is close to life and can incorporate the various clues and stages that occur in real--not textbook--conversions. The *Summa Theologiae* itself is not a system as described above. It is a magnificent synthesis. It incorporates a vast amount of data from many different sources. It has governing and explanatory principles, but these are not arranged in a series of deductions.

---

[10] *Ibid*., p. 78.

[11] *Ibid*., p. 78.

III. A Pluralism of Systems or of Syntheses?

Now we can consider the question of a pluralism of systems. Is it possible for opposed systems to converge analogously toward a more adequate truth? Using the previously established definition of system, the answer must be "no." Systems that are opposed must be opposed on their fundamental governing principles and no analogy can circumvent the principle of contradiction. Spinoza and Hume cannot be considered as converging toward any more adequate position.

Now it is possible to have a variety of syntheses which indeed can converge toward a more adequate understanding of truth. A condition is that the pervasive principles be not in direct opposition. The emphasis, the aspects, the interrelationships, the selection of even the individual arguments can be different. Yet a variety of differing, not "opposed", syntheses is possible and can indeed converge toward a more adequate understanding.

Let me illustrate this from a quite different subject matter. Suppose there is an Alpine valley surrounded by lofty peaks, adorned with trees with burnished green leaves. It is spring and the valley is bright with clusters of fresh flowers. Now we ask five artists to paint a picture of this scene. Suppose four of them produce uniquely different impressions of the valley--some giving it a fairy-like atmosphere, others exploring its lush exuberance of leaves and flowers, of brilliant peaks and lovely rock. These we could recognize as individual "syntheses" of the natural beauties of the real valley. Now the fifth artist presents a picture of a flat desert, dark sky, and one cactus plant. This could in no way be considered as another synthesis of the beauties of this scene.

The epistemologies of Jacques Maritain, Etienne Gilson, Joseph Owens and other Thomists have produced personal syntheses in epistemology. They supplement, explain and, to some extent, correct each other. No one can say that one synthesis is absolutely right and another absolutely wrong. But, now put Hume alongside these and we have an analogy with the desert scene. The kind of world Hume imagines doesn't exist and he can in no way be considered to converge with the Thomists to a more adequate truth. I will return to the problem of this "convergence" later.

IV. History and Philosophy[12]

The relation of history to theology and formulation/reformulation of Catholic dogma is quite different from the relationship of history to the development of philosophy. Because of my concentration on philosophy in this paper, I will say nothing of the first relationship.

There are, of course, contingencies which modify the actual history of philosophy. If the Aristotelian *corpus* had not been available in Latin in the thirteenth century, the whole course of medieval philosophy would have been changed. And what if the lightning that hit the tower where Thomas, the child, was sleeping had killed Thomas instead of his sister? What if Kant had died before reading Hume? There probably would have been no *Critique of Pure Reason* and Thomas Reid's devastating attack on Hume might have had the central position in anti-Humean criticism.

There are, however, some truths and facts that, once discovered, stand free on their own evidence, free not only of the history of their own discovery, but also of all history and often need no reformulation. Probably the Pythagoreans discovered that the square on the hypotenuse of a right triangle is equal to the sum of the squares on the other two sides. Certainly Euclid put it into his own geometry. Why did the Egyptians, who used triangles in surveying, not work out geometry? No matter. The truth of geometry can be taught to an

---

[12] *Ibid.*, pp. 204-208.

American schoolboy as well as to a Tibetan monk, or as well to a Scottish lass as to an African chief--and without any "reformulation."

The case is the same with many scientific facts. It is an atemporal fact that electricity can produce illumination (though the explanation of the fact may be uncertain). This scientific fact is now atemporal, quite independent of its history, and super-cultural, being taught around the world without cultural modifications. Modern mathematical physics, though the product of a long cultural history going back at least to Classical Greece, is now accepted simply in its own right throughout all the cultures of the world.

In a somewhat different way, a similar case can be made for philosophical problems, insights and truths. There are many such insights and truths in St. Thomas that can be extracted from the milieu of the thirteenth century and from the Thomistic text itself. Thus, when the principles of act and potency have been established, or the primacy of the act of existence recognized, or the problem of universals has been solved, those truths become atemporal and culturally free. They are then available for use in a great variety of philosophical investigations and syntheses with a minimum of reformulation.

Of course there are areas of human concern in which philosophical revision and new developments are necessary. These areas include health care, economics, politics and social and cultural institutions. The secondary and tertiary principles of the Natural Law generally contain a contingent element. A brief simple instance may illustrate this. St. Thomas says that a lawyer commits a sin if he defends a client he knows to be guilty.[13] But in the judicial systems developed in England and the United States, wherein an accused person is deemed innocent until convicted in a court of

[13] St. Thomas, *Summa Theologiae*, II-II, q. 70, a. 3.

law, lawyers must undertake to defend clients regardless of their own knowledge of their clients' guilt. This is necessary to maintain even justice throughout the system.[14]

The case of error is quite different. The emergence of error requires an historical explanation, including human decisions and non-rational pressures. Kant's personal decision to accept the basic conclusions of Hume is necessary to understand the origin and content of the *Critique of Pure Reason*. Karl Marx cannot be understood except in the context of the economical developments of his time. To understand Kant's *Critique of Pure Reason*, one must have a thorough knowledge of Hume.

V. Thomism and Thomisms

It is obvious that the Thomistic revival has generated a pluralism of Thomisms. It may clarify matters if we can identify and classify those versions of Thomism which constitute that pluralism. For this purpose I will use the following designations: 1. The Neo-Thomistic Movement; 2. Authentic Thomism; 3. Thomistical Development. This last category I will later subdivide.

First, then, we can refer to McCool's description of the initial program of the Neo-Thomistic Movement as set forth and derived from *Aeterni Patris*. As McCool has shown, many of the assumptions and expectancies

---

[14] "This argument seems questionable, at least from the point of view of the ethics of the English and Scottish legal profession. For it seems to assume that a lawyer always wills that his client's case succeeds. In actual fact, a lawyer's duty is to ensure that a client's case (whether 'unjust' or not) secures a proper hearing: he would be usurping the judge's function if he decided the justice of his client's case for himself." Footnote to text cited in note 13 (*Summa Theologiae*, Blackfriars edition, New York: McGraw Hill, 1975).

of that initial program were mistaken and illusory, so we can say that that initial program as such no longer exists. Yet, it had an enormous impact on Catholic thought and education and gave rise to the very scholarship that brought about its "demise." Since all these writings in this initial stage have a certain similarity, I will simply call them Neo-Thomistic.

In the second classification we have, first of all, the Thomism of St. Thomas himself as contained in his own writings. Here also I place the work of those who, with utmost loyalty to the original and with careful historical scholarship, have established the meaning of the Thomistic texts and explained their doctrines. Here I would include the work of men like Gilson and Ignatius Eschmann. All this I would call "authentic Thomism."

However, it would be unreasonable to freeze Thomism as it was when St. Thomas died. St. Thomas' thought was constantly growing. He was anxious to get better translations and more sources. As soon as a translation of a relevant writing appeared St. Thomas immediately began to use it. If he had lived ten or fifteen more years he would not simply have repeated himself. So we have a vast Thomistical literature which I will now attempt to classify.

In this Thomistical literature we can distinguish at least four types. First, there exist writings which profess to explain authentic Thomism but fail to do so. Thus McCool points out that Cajetan's three degrees of abstraction and his doctrine on the *esse* of accidents show that he misunderstood St. Thomas.[15] Second, one finds creative developments of St. Thomas' doctrines. Thus, John Finnis pointed out that St. Thomas himself never worked out a complete theory of Natural Law.[16] In Vitoria, Vásquez, Bellarmine, etc., we find an

---

[15] McCool, *op. cit.*, pp. 155-156.

[16] John Finnis, *Natural Law and Natural Rights* (Oxford: Clarendon Press, 1980), pp. 45-57.

extensive and brilliant development of Natural Law as it applies to law and politics. Third, Thomistical literature contains work that applies Thomistic principles to entirely new areas of culture. Another example could be Jacques Maritain's brilliant work in aesthetics. Fourth, we find attempts to develop a separate philosophy. From the very beginning of the Thomistic revival an effort was made to extricate St. Thomas' philosophical positions from their theological context and to present them in a series of disciplines (e.g., Metaphysics, Rational Psychology, Epistemology, etc.). What of the possibility of succeeding in such an effort?

Gilson has shown, according to McCool,[17] that St. Thomas worked out his own philosophy in and through his theologizing. Thus authentic Thomism exists in a theological context and as part of St. Thomas' theology. Yet, if St. Thomas' philosophical positions are true and truly philosophical, they should be able to be extracted from theology and set forth simply as philosophical. A simple example can be found in *Summa Theologiae*, I-II, q. 97 wherein St. Thomas discusses "change in law." The context of this Question can easily be reformulated in a modern manner and favorably compared with, for example, Lon Fuller's discussion of the same topic.[18]

Moreover, the continued development of a separated Thomistic (or Thomistical) philosophy is necessary if this philosophy is to enable us to enter into modern discussions and influence modern thought. St. Thomas himself recognized in principle the necessity of such a separation when he wrote:

> In the second place, it is difficult [to refute others' errors] because some of them, such as the Mohamedans and the pagans, do not agree with us in

---

[17] McCool, *op. cit.*, pp. 165-170.

[18] Lon Fuller, *The Morality of Law* (Revised Edition, New Haven: Yale University Press, 1964), pp. 79-81.

> accepting the authority of any Scripture, by which they may be convinced of their error. Thus, against the Jews we are able to argue by means of the Old Testament, while against heretics we are able to argue by means of the New Testament. But the Mohammedans and the pagans accept neither the one nor the other. *We must, therefore, have recourse to the natural reason, to which all men are forced to give their assent.* [19] [emphasis added]

Of course, no one knows how St. Thomas would teach, say, metaphysics in a modern university. Joseph Owens has speculated interestingly on St. Thomas' ordering of such a presentation.[20] But the re-presentation of separated Thomistic philosophy in different disciplines requires the creative work of modern Thomists. All such presentations must at least remain Thomistical, not Thomistic in the strong sense.

I digress here in order to explain why I do not list Transcendental Thomism either as authentic Thomism or as Thomistical. McCool includes[21] Transcendental Thomism among the *internal* developments of Thomism. It seems to me that it was rather an imposition from the outside, the imposition of a methodology utterly foreign to Thomism. The author of the movement was Joseph Maréchal, S.J., who set forth his position in a brilliant series of volumes entitled, *Le Point de départ de la métaphysique*, especially in Cahier V. [22] His project was to use Kant's own methodology, the transcendental method, to arrive at Thomistic conclusions,[23] thus

---

[19] *Contra Gentiles*, Bk. 1, Chap. 2.

[20] "A Note on the Approach to Metaphysics", *The New Scholasticism*, 28 (1954), 454-476.

[21] McCool, *op. cit.*, pp. 82-113.

[22] *Le Point de départ de la métaphysique: Lecons sur le développement historique et théorique du problème de la connaissance.* 5 vols. (Paris: Desceé De Brouwer, 1944-1949).

[23] McCool, *op. cit.*, pp. 105-106.

making Thomism a simulacrum of a modern critical philosophy. A group of brilliant thinkers enthusiastically followed Maréchal's lead.[24] These men have made insightful permanent contributions to philosophy,[25] but the project itself has not succeeded,[26] and, if pushed to its logical conclusion, will finally end in skepticism.[27]

In accepting the Kantian methodology Transcendental Thomists implicitly accepted Kant's formulation of the problem of human knowledge. Kant's formulation in turn rested on Hume's conclusion that no necessity, no universality and no intelligibility

---

[24] J. de Fries, S.J., A. Marc, S.J., G. Isaye, E. Coreth, S.J., Karl Rahner, S.J., B. J. F. Lonergan, S.J., A. Gregoire, J. B. Lotz, et. al.

[25] Cf. Etienne Gilson, *The Unity of Philosophical Experience* (New York: Charles Scribner's Sons, 1947), p. 301: "On the other side, all those subtle shades of thoughts which qualify the principles of a philosopher, soften their rigidity and allow them to do justice to the complexity of concrete facts, are not only part and parcel of his own doctrine, but are often the only part of it that will survive the death of the system. We may wholly disagree with Hegel, or with Comte, but nobody can read their encyclopedias without finding there an inexhaustible source of partial truths and of acute observations. Each particular philosophy is, therefore, a co-ordination of self and mutually limitating principles which defines an individual outlook on the fullness of reality. What Gilson says here of all great philosophers applies to the Transcendental Thomists, especially since they retain many basic Thomistic insights.

[26] I have made this point in my critique of Coreth. See "Transcendental Thomism: A Critical Assessment" in *One Hundred Years of Thomism: Aeterni Patris and Afterwards, A Symposium*, Victor B. Brezik, C.S.B., ed., (Houston: Center for Thomistic Studies, University of St. Thomas, 1981).

[27] Thomas Sheehan's analysis of Karl Rahner's metaphysics shows how close to agnosticism Rahner came at the very beginning of his career. *Karl Rahner, The Philosophical Foundation* (Athens: Ohio University Press, 1987), pp. 139-172.

can be found in experience. The Humean presupposition infects the whole of the *Critique of Pure Reason* and consideration of it reveals the transcendental method as a *tour-de-force* to solve a false problem.

The Humean presupposition is why the Transcendental Thomists insist so strongly that there is no intuition of being in experience. It is also why Joseph Donceel wrote:

> That is why metaphysics is *a priori*, virtually inborn in us, not derived from sense experience, exactly as the soul is ontologically prior to the body and not derived from it. Yet we would never know any metaphysics if we had no sense knowledge, exactly as our soul cannot operate without our body. Hence metaphysics is *virtually* inborn in us. We become aware of it only *in* and *through* sense knowledge, although it does not come *from* sense knowledge.[28]

I do not see how a system starting at a radically different starting point, with radically different assumptions, employing a radically different methodology and drawing from it pseudo-Thomistic conclusions can be called a Thomism of *any* kind (certainly not Thomism in the strong sense).[29]

---

[28] Donceel, *op. cit.*, p. 8.

[29] One of the most troubling things about Transcendental Thomists is their explicit hermeneutic methods in dealing with the Thomistic texts. They do not attempt to determine scientifically or with careful exegesis the meaning of the Thomistic texts. Rather they use the Thomistic texts to reveal things beyond the text itself, things "not said" in the text. This makes it possible for them to support their positions with quotations from St. Thomas in support of very un-Thomistic ideas. See Karl Rahner's explanation of his exegetic method in *Spirit in the World*, William Dych, S.J., tr. (New York: Herder and Herder, 1968), pp. l-liii.

I now return to the main discussion. In order to make these Thomistical presentations relevant to modern culture, one must not only have an accurate knowledge of authentic Thomism and the subsequent Thomistical developments but also a basic understanding of modern philosophy. Obviously this task requires intellectual ability of the highest order as well as meticulous scholarship.

VI. Divergence and Convergence in Thomistical Developments[30]

In any living philosophical tradition, it is inevitable, as McCool establishes for Thomism, that pluralism results. As brilliant minds took Thomism seriously and engaged in relevant research as well as philosophical reflection, being of differing backgrounds, education and abilities, they worked out different re-presentations of Thomism. Since in philosophy, as Rousselot showed, insight (*intellectus*) is central and since there are no logical rules or objective relevant experiments for gaining an insight, it is impossible that any group of intelligent scholars should achieve exactly the same insights.

This is true of all Thomistical developments but it is especially evident in the elaboration of a separated philosophy, in its organization into distinct disciplines and in its presentation in different textbooks. Some of these texts are so divergent that they are incompatible with any Thomistic tradition. Such is the case of Donceel's English edition of Coreth's *Metaphysik*.[31]

---

[30] McCool, *op. cit.*, pp. 212-219.

[31] Emerich Coreth, *Metaphysics*, English edition by Joseph Donceel, S.J., (New York: Herder and Herder, 1968). The title page is misleading since it simply calls this work an "English edition." As Father Donceel says in his introduction, large sections (some, in my opinion, were important) have been omitted.

On the other hand, many of the textbooks produced in the last fifty years are not only mutually compatible but also converge towards a fuller and more adequate understanding. This is the case with Joseph Owens' *An Elementary Christian Metaphysics,* [32] Klubertanz's *An Introduction to Metaphysics,* [33] Vaske's *An Introduction to Metaphysics,*[34] and Joseph Owens' *An Interpretation of Existence.*[35]

VII. The Construction of a Thomistical Epistemology

What I have said above about a separated philosophy can be illustrated by considering how one might work out a modern Thomistical epistemology.

Discovering the elements in St. Thomas for the construction of a modern-type epistemology is much more difficult than for any other branch of philosophy. Throughout McCool's book, when he talks of knowledge in St. Thomas he usually talks of St. Thomas' "metaphysics of knowledge." St. Thomas generally dealt with knowledge in a theological-metaphysical way, though sometimes his treatment could be called psychological. This, in effect, makes epistemology a chapter in metaphysics. Joseph Owens

---

The original German edition had 584 pages; the "English edition" has 224 pages.

[32] Joseph Owens, C.Ss.R., *An Elementary Christian Metaphysics* (Houston: Center for Thomistic Studies, 1985).

[33] George Klubertanz, *Introduction to the Philosophy of Being* (New York: Appleton-Century-Crofts, 1963).

[34] Martin O. Vaske, *An Introduction to Metaphysics* (New York: McGraw Hill, 1963).

[35] Joseph Owens, C.Ss.R., *An Interpretation of Existence* (Houston: Center for Thomistic Studies, 1985).

maintains this location of epistemology in his *An Elementary Christian Metaphysics.*[36]

Neither St. Thomas nor his immediate and close followers ever conceived a philosophical discipline which would have knowledge not only for its material object but also for its formal object--a discipline which, though not totally independent of metaphysics and psychology, would, nonetheless, constitute a self-standing and separate discipline. Modern philosophy has been beset with epistemological problems ever since Descartes began by separating "thinking" from all direct contact with external reality and made it necessary for a realist to establish the validity of our ideas of the material world. As the subsequent history shows and as Gilson demonstrated in his *Réalisme thomiste et critique de la connaisance*,[37] any philosophy that excludes direct realism inevitably leads to skepticism or idealism. Thus, despite Descartes' complicated effort to prove that the world corresponds in some way to our sense ideas, British Empiricism reached its *reductio ad absurdum* in Hume. Maréchal's effort to reach Thomistic realism by using a Kantian methodology also failed and for the same reason.

St. Thomas took direct realism for granted. However valid that realistic position is, we cannot do that today. Both Neo-Thomism and modern philosophy in general need an epistemology that vindicates direct realism and rejects both rationalism and empiricism. No wholly satisfactory Thomistical epistemology of this sort has as yet been constructed.

---

[36] Joseph Owens, C.Ss.R., *An Elementary Christian Metaphysics*, pp. 131-191.

[37] Paris: Vrin, 1947.

VIII. The Basic Elements of a Thomistical Epistemology

The basic elements may be listed as follows. First, a Thomistical epistemology is a solid direct realism.[38] There are many descriptions of direct realism, but it can be summarized in three propositions: there is a really existing material world, extended in space, three-dimensional and existing independently of the knowledge and wishes of human beings; we can gain considerable objective knowledge of this world; the previous two propositions cannot be demonstrated and need not be demonstrated since they are self-evident in human experience.

Second, a Thomistical epistemology recognizes a radical distinction between sense experience and intellectual knowledge, without separating their activities. The complete separation of "thinking" [Descartes] from sense knowledge of the world is just as disastrous as a reduction of intelligence to sense knowledge.

Third, a Thomistical epistemology contains a careful epistemological analysis of *perception* defined as the combined activity of sense and intellectual knowledge leading to a perceptual judgment about an object or objects presented here and now in sense experience. A thorough epistemological analysis of sense knowledge that respects the character of the individual senses and their interrelationships has not been made. Most Thomistical epistemologies deal with the senses in a formalistic univocal way. The empirical sensist analysis

---

38 "A Phenomenological Approach to Realism," *An Etienne Gilson Tribute* (Milwaukee: Marquette University Press, 1959), pp. 68-85; "Basis of Philosophical Realism Re-examined," *The New Scholasticism*, 56 (1982), 1-29; *Theory of Knowledge*, R. J. Henle, S.J. (Chicago: Loyola University Press, 1983), 80-128; "Schopenhauer and Direct Realism," *Review of Metaphysics*, 46 (1992), 125-140.

has failed. The intuitive function of intellect [*intellectus*; Rousselot] as it penetrates and pervades sense knowledge must be stressed. The existential character of perceptual judgment as well as its implicit expression of the contingent necessities of being must be maintained (as against all forms of a priorism [Kant, Maréchal, Lonergan, et. al.]).

Fourth, a Thomistical epistemology acknowledges as essential elements the intentionality of intellectual knowledge and the function of the concept as a *formal* sign [John of St. Thomas].

Fifth, a Thomistical epistemology recognizes the ordination of the speculative intellect to truth and that the formal object of the intellect is being.

Sixth, one of the basic problems that pervades the whole of modern culture is the question of the range and validity of the different disciplines and methodologies that constitute a great part of our culture. The Cartesian ideal of a single demonstrative discipline using a single univocal methodology keeps reappearing. And, of course, the methodological imperialism about which Gilson has written so well. Does evolutionary science prove that human beings are simply animals, a bit more clever and perhaps more aggressive than other species? Did Logical Positivism eliminate metaphysics? Etc., etc. A modern epistemological analysis of the methodologies, the subject matter and the range of certitude in the major types of modern organized knowledge is essential, as St. Thomas did for mathematics in *Expositio super librum Boëthius de Trinitate* and Maritain did (despite the criticism of Gilson and McCool) with some success in *The Degrees of Knowledge*.[39]

---

[39] I have dealt with this problem in *Theory of Knowledge*, pp. 288-336. See also "Science and the Humanities," *Thought*, 35 (1960), 513-536; "Philosophical Method and the Cultural Crisis of Our Times," *ACTAS Segundo Congreso Extraordinario*

Seventh, this modern Thomistical epistemology, though a distinct discipline, must be compatible with and even imply a metaphysics of existential being [St. Thomas; Gilson].

Finally, it should be clear that this epistemology should be a synthesis (as indicated by the sample elements listed above) rather than a system. It also must be organized and taught, not in the rigid systematic abstract method of the early "Thomistic" manuals. It should be taught in an inductive and experiential manner which can reveal even to undergraduates the convincing evidence for the various essential positions.

I do not think that the importance of such a Thomistical epistemology for modern culture and especially for Catholic liberal education can be overemphasized. The epistemological confusion in the culture of our age is one of the most detrimental, and even dangerous, aspects of that culture.

IX. Current and Future Thomism

The original official protocol of the Thomistic Movement may, as McCool shows, have turned out to be mistaken and nonfeasible. Thomism thereby has lost its officially approved and promoted position and its consequent wide acceptance in colleges, universities and seminaries. But "demise" is too strong a word, even for the original protocol. After all, that protocol itself contributed significantly to its transformation by internal development.

And so I do not think that, because of its development to pluralism, Thomism will now become an object of historical study only. It is living philosophy and is still being explored in its original authentic form and in its Thomistical developments. Associations, congresses, publications and centers are actively

---

Interamericano de Filosofia, 22-26 Julio 1961, Costa Rica: Imprenta Nacional, 1962, pp. 257-260.

promoting it as do other relevant groups like the various Maritain Societies around the world which, for example, continue to guide the Christian democratic movement in Latin America.

Thomistical courses should also be prepared and promoted. Metaphysics, epistemology and ethics should be part of the core curriculum in any liberal education, especially in Catholic liberal education. After all, Thomistical philosophy is the only current living philosophy that can justify itself as a separated philosophy and yet is compatible with and supportive of religious faith.

X. Conclusion

I repeat what I said in the beginning. This is a brilliant book. Father McCool has made a significant contribution to the understanding of modern Thomism and has given direction and inspiration for its future development. I agree with the last paragraph in his introduction:

> The tradition of St. Thomas is a living and evolving one. Its internal conflicts are not yet over, and all its contributions to contemporary thought have not yet been made. It is our hope that this study of its internal evolution in our century may lead to a greater understanding of the nature and value of its contribution to contemporary thought.[40]

---

40 McCool, *op. cit.*, p. 3.

TRUTH, REALISM AND PHILOSOPHICAL PLURALISM

Marc F. Griesbach

In today's world, to mention philosophy at all inevitably raises the question: which philosophy? At whatever point one takes up an examination of the history of philosophy, there will be a plurality of systems to consider. And if one were to limit his consideration to a single philosophical "family," such as Platonists or Aristotelians or Thomists, a similar situation would pertain. It is hardly surprising, then, that Etienne Gilson's careful inquiry into the nature of medieval ("scholastic") philosophy revealed a plurality of quite distinct "Christian" philosophies,[1] nor that Thomistic philosophies in the sixteenth century should all be found to deviate significantly from the thirteenth century philosophy of St. Thomas Aquinas. Even less surprising is it that the effort to restore Thomism in the nineteenth and twentieth centuries has resulted in multiple claims to "authenticity" made on behalf of an assortment of "Thomisms," both "transcendental" and "non-transcendental."[2]

---

[1] See Etienne Gilson, *The Spirit of Mediaeval Philosophy*, transl. A. H. C. Downes (New York: Scribners, 1940); *History of Christian Philosophy in the Middle Ages*, (Random House: New York, 1955), especially pp. 544-45; *The Philosopher and Theology*, trans. by Cecile Gilson (New York: Random House, 1962), pp. 92 ff.

[2] "Transcendental" Thomisms are included here in my enumeration, not because I concede the logical coherence of such a position (I do not), but because such a doctrine will be at issue in the topic which will be our principal concern. For an extensive treatment of this particular point, see Etienne Gilson, *Thomist Realism & Critique of Knowledge*, transl. Mark A. Wauck (San Francisco, CA: Ignatius Press: 1983), especially chs. 5 and 6.

Such a "pluralistic" state of affairs does not, of course, prevent one from legitimately speaking of a single philosophic enterprise comparable to such scientific enterprises as those of chemistry, physics, biology, astronomy, etc. Yet even here, it may seem much more appropriate to say that students of physics, for example, are seeking to acquire one and the same body of knowledge than to say something similar of philosophy majors in various universities offering such a program of studies. To go further, acknowledging, in this day and age, that one is seriously engaged in the pursuit of wisdom, a "system" of philosophic knowledge universally true for all time, a "perennial philosophy," would surely require the expenditure of a considerable amount of modern academic respectability.

And yet, many of us still remember that philosophy had once been regarded as the pursuit of just such a wisdom, which, though admittedly beyond human achievement in its perfection, was considered a most worthy object of our striving. Moreover, there seems to be good reason to believe that, well into what we now call modernity, every serious student of philosophy understood himself to be engaged, within the limits of his ability, in just such a search for a perennial wisdom.

Even today, persons philosophically educated in the Catholic tradition know that Pope Leo XIII, in his famous encyclical[3] of 1879, *Aeterni Patris* (On the Restoration of Christian Philosophy), called for the restoration of the "perennial philosophy" of St. Thomas Aquinas, especially to provide the philosophic basis for the seminary training of priests. This influential

---

[3] English translations of the encyclical can be found in such works as *The Wisdom of Catholicism*, ed. Anton C. Pegis (New York: Random House, 1949); Jacques Maritain, *St. Thomas Aquinas*, transl. J. F. Scanlon (London: Sheed and Ward, 1931); *The Church Speaks to the Modern World: The Social Teachings of Leo XIII*, ed. Etienne Gilson (New York: Doubleday Image, 1954).

document, often referred to as the "magna charta of the Thomistic revival," was to play an important role in Catholic intellectual life for roughly three quarters of a century. Though Catholic philosophers were also urged, in that document, to keep their minds open to ongoing scientific and philosophic achievements, and to incorporate into the essentially perennial philosophy whatever modifications the truths contained in them required, this Thomistic mode of thought was declared to be the proper way to philosophize.

There followed a period of great promise for the widespread acceptance of this mode of philosophizing, unforgettable by those of us who were privileged at the time to be studying under such masters of the thought of Thomas Aquinas as Gilson, Maritain, Pegis, Eschmann and Simon. Then, almost as quickly, while we were in the process of achieving our own philosophic maturity, those high hopes seemed, suddenly, to evaporate. Many, even of those whom we had regarded as our "fellow Thomists," seemed to have lost their nerve, and joined in the chorus celebrating the "death of Thomism." All too soon, the revival of the "perennial philosophy" was over.

Enough, though, of such nostalgic reminiscing. The question I propose to consider here is whether the very notion of a perennial philosophy has actually been shown to be untenable, once and for all, and a doctrine of pluralism-in-principle firmly and legitimately established in its place. More specifically, we will be concerned with the relationship of this issue to the traditional epistemological doctrines of direct realism and truth.

Recently (in 1989), Gerald A. McCool, S.J., of Fordham University, published a scholarly study analyzing, for the most part, the thought of four significant figures in Thomistic thought during the past century, and concluding, somewhat abruptly, in my judgment, that authentic Thomistic thought is (and can only be) intrinsically pluralistic. That book, *From Unity*

To Pluralism,[4] only reinforces the need, already felt by some of us schooled in the Thomistic tradition, and dissatisfied with currently prevailing philosophic attitudes, to give further careful thought to the implications of this new philosophic "modesty," with respect to the traditional direct realist view of man's cognitive capability.

If I understand him correctly, Father McCool, in the closing pages of *From Unity to Pluralism*, seems to be finally persuaded that the plurality of interpretations of St. Thomas' essential thought is a necessary consequence of authentic Thomistic epistemological principles. He seems to have been finally persuaded by a series of arguments to the effect that the judgments which constitute any human philosophic system provide, at best, only a finite "partial truth," due to the analogical character of human concepts. Accordingly, a plurality of such tenable-though-individually-deficient philosophical systems is not only possible, but much to be desired--even logically warranted.

The philosophical and theological journey along which Father McCool takes us in coming to this position, particularly in the final chapter of his book, is too long and involved for us to travel with him here, making every stop with him along the way, especially

---

[4] *Op. cit.* (New York: Fordham University Press, 1989). After a first chapter devoted to a detailed account of how *Aeterni Patris* came to be written as the cumulation of a 19th century revival of medieval scholasticism, notably in Italy, the author devotes the next seven of the remaining eight chapters primarily to a consideration of the epistemological thought of Pièrre Rouselot, Joseph Maréchal, S.J., Jacques Maritain and Etienne Gilson. He also gives considerable attention to the philosophical and theological controversies concerning the ultimate end of man and to the notion of "Christian philosophy." A final chapter, contrasting the thought of such prominent European Jesuits as Eric Przywara, Henri Brouillard and Jean Marie LeBlond with that of Maritain and Gilson, brings Father McCool to his "pluralist" conclusion.

inasmuch as for him the issue throughout is framed largely in terms of competing "old" and "new" theologies.[5] In contrast, my chief concern in the matters I intend to pursue here has to do with the philosophical implications of such a supposedly benign and principled pluralism.

Nevertheless, it seems important that we take note of the principal steps by which Father McCool finally concludes that Thomistic principles themselves lead to philosophic pluralism. To begin with, he seems to have become convinced that Pope Leo's encyclical, *Aeterni Patris*, is flawed on at least two counts: first, as mistaking the multiple "scholasticisms" of the 19th century as the authentic doctrine of St. Thomas Aquinas; and second, as assuming (falsely) that Thomism was capable of assimilating all truths of all time. The latter assumption, he contends, has long since been discredited; the former is now thoroughly untenable, because of the careful research of the impeccably-accredited Thomist, Etienne Gilson.[6]

It was the rather novel application of St. Thomas' doctrine of analogy, however, which Father McCool seems to have accepted from the group of European Jesuits to which we have referred above, that formed the basis for his pluralist position. As is well known, we are able, according to the Thomistic doctrine of analogy to proceed from a knowledge of a human attribute, e.g., man's goodness, to a knowledge that God also is good: but whereas man's finite goodness is proportional to his finite mode of being, God's goodness is infinite, as proportional to His infinite being, which is, of course,

---

[5] McCool, *From Unity to Pluralism.* See especially ch. 9.

[6] Gilson, along with his confrere, Jacques Maritain, remained firmly committed to the principal teachings of *Aeterni Patris*, in particular to the preeminence of Thomism as the *philosophia perennis*, and published his own English edition of the encyclical together with the Pope's other social writings (cited above in note 3).

beyond our ability to comprehend. Proceeding, according to this new and, to say the least, unusual use of analogy, from God's infinite knowledge to man's finite knowledge, it is argued that absolute truth can be attributed only to God, whereas all human truth, as finite, necessarily falls short of being "absolute."[7]

All human concepts, it is then argued, are defective as falling short of absolute truth. Hence, again according to this line of reasoning, the same must be said of all the judgments which constitute any given "system of philosophy." Given the multiplicity of such "systems of philosophy," each of them able to represent reality only "analogously," that is to say, "inadequately," philosophical pluralism seems destined to be the ultimate in human wisdom.[8]

This brief summary of the thinking upon which Father McCool seems to rely does not, of course, take

---

[7] *Ibid.* See especially pp. 214-219. Although it is commonly understood that, for St. Thomas, the doctrine of analogy clearly applies to the relationship between our human knowledge and the Divine Knowledge, it does so in the sense that God's knowledge so far transcends our knowledge that we can know that God knows infinitely, but we cannot understand what His Infinite Knowledge is. This does not mean, however, that from the fact that God's knowledge (alone) is infinite (perfect), we can conclude that all human knowledge, as finite, is deficient in the sense that we necessarily fall short of achieving any (unqualified) truth.

[8] *Ibid.*, p. 216. The argument offered here: that "No finite mind can encompass the whole reality of God's ineffable infinity," though unexceptionable for any follower of St. Thomas, is insufficient to make the point intended. See below (also on p. 216): "Thus each speculative system gives expression to the dynamic balance between man and God *in its own unique and analogous way* (emphasis added). From this it follows that the abiding fullness of the Church's religious truth cannot be expressed through a single exclusive system. It finds its expression only through the convergence of the polarity of the different systems of her philosophy and theology, each of which can express an aspect of the total truth which the others cannot."

into account the nuances that characterize the individual views of the several contributors he cites. Pierre Rousselot, who died in his youth on a World War I battlefield, had written an impressive book emphasizing the "intellectualism" of St. Thomas, (the role of *intellectus* in his epistemology) as against his rationalism, especially his doctrine of human conceptual knowledge. Whether or not he would have subscribed to such a pluralist conclusion must remain in question. The rector of the Catholic Institute of Toulouse is described by Father McCool, somewhat ambiguously, as believing that "in an analogous way, Thomism comes closer to absolute truth than any other system."

Interestingly, the author of *From Unity To Pluralism* raises no objection to the peculiar use made of the Thomistic doctrine of analogy by those whom he sees fit to follow along the path to the "New Theology." He finally and decisively commits himself to the "future," which he judged "lay with the 'New Theologians'," only in the closing pages of his book,[9] and only after he had given an ostensibly fair hearing to the "conservative Thomists," including Maritain and Gilson. He is now fully prepared to consign the notion of a Thomistic "*philosophia perennis*," once and for all, to the scrapheap of history.[10]

Apparently not yet quite content, Father McCool seeks to draw further support for his position from the Second Vatican Council's *Decree on Priestly Formation.* Having accurately quoted a crucial sentence: "Basing

---

[9] *Ibid.* See pp. 224-30.

[10] *Ibid.* A favored, apparently less "conservative," remnant of "Thomism" was, nonetheless destined to survive among the "new theologians." See p. 225: "As the history of theology after Vatican II was to show, the future lay with the 'new theologians,' and the form of Thomism which LeBlond used to vindicate the place of history and pluralism in theology is the form of Thomism which survived the demise of the Neo-Thomist movement in the theologies of Rahner and Lonergan."

themselves on a philosophic heritage which is perennially valid, students should also be conversant with contemporary philosophical investigations, especially those exercising special influence in their own country, and with recent scientific progress," he interprets this as a vindication of the "new theologians."[11]

Unless I am mistaken, a great deal is at stake philosophically, not just in Father McCool's proclamation in his book, *From Unity To Pluralism*, but in today's broader claim of many of our modern colleagues that pluralism in philosophy is not just a fact of the current intellectual situation, but the only reasonable position for any "enlightened" philosopher to take.[12]

At first sight, nothing would appear to be more benign than a view of human knowledge which advises us to be content to look upon the various contributions of philosophers as so many more or less interesting, creative, challenging attempts to deal with those aspects of reality which are beyond the limits of a genuine "scientific" understanding. Lacking a method by which to test the truth of conflicting philosophical notions, a proper humility would seem to be only appropriate.

---

[11] *Ibid.*, p. 229. Even a cursory rereading of *Aeterni Patris* will show that this only reiterates what Pope Leo's document already clearly stated.

[12] Deserving our particular attention in this regard is an important 1992 publication of the American Maritain Association entitled: *The Future of Thomism.* Edited by Deal W. Hudson and Dennis Wm. Moran, this volume is comprised of more than a score of essays by leading Thomist scholars, including a brief preface and an article entitled: "Is Thomas's Way Still Viable?" by Father McCool. In summarizing many of the points he had made in the book we have been considering, the author pays considerable respect to the leading figures of the Neo-Thomist movement which he now declares moribund, while holding out considerable hope for the future of the "broader Thomist Tradition."

Here in our philosophizing, if anywhere, a generous attempt to accommodate a multiplicity of views would seem to be the proper attitude. Where, after all, would such virtues as open-mindedness and tolerance seem more appropriate?

Of course many of us are aware that it does not follow from the fact that philosophical issues are not, like scientific hypotheses, susceptible to testing by way of controlled experiment, that they are, on that ground, to be relegated to a realm where we must learn to be content with a plurality of disparate views, no one of which is unqualifiedly true.[13] Moreover, it should be obvious to all of us that to be open-minded and tolerant requires of us only that we give respectful consideration to persons holding views seemingly opposed to our own, not that we regard those views as true (or even as partial truths, along with our own.)[14] Perhaps our very discomfort with such words as "truth" used in this context is a sign that the attempt to delineate this

---

[13] I am reminded here of what A. J. Ayer once wrote in a somewhat similar context concerning the positivistic downgrading of what they call "metaphysical" statements. "No, it does not follow. Or rather, it does not follow unless you make it follow." *Logical Positivism*, ed. A. J. Ayer (Glencoe, IL: The Free Press, 1959), Introduction, p. 16.

[14] See Jacques Maritain, *On the Uses of Philosophy* (New York: Atheneum, 1965), pp. 22-24. In answer to those who think that tolerance requires one to be a relativist or a skeptic, Maritain writes as follows: "There is real and genuine tolerance only when a man is firmly and absolutely convinced of a truth, or of what he holds to be a truth, and when he at the same time recognizes the right of those who deny this truth to exist, and to contradict him, and to speak their own mind, not because they are free from truth but because they seek truth in their own way, and because he respects in them human nature and human dignity and those very resources and living springs of the intellect and of conscience which make them potentially capable (sic) of attaining the truth he loves, if someday they happen to see it."

"pluralist" position has stretched our more traditional vocabulary beyond its proper limits.

It is also one thing, moreover, to be convinced of the value of a thorough historical knowledge of the vast scope of philosophical thought from ancient times to the present day; it is something quite different to adopt the attitude of philosophical pluralism-in-principle. By all means, let us "keep our intellectual horizons always open and receptive to the experiences of men and the findings of fellow philosophers throughout the ages," as our old friend, James Collins, once put it.[15] There is merit, assuredly, in the current striving for a diversity of views in many of our philosophy departments, provided, I would insist, that we maintain channels of communication sufficient to guarantee their frequent clashing in the continuing search of all for truth.[16] (Here the word seems altogether appropriate again.)

Let us, however, distinguish sharply between an abiding concern for truth by way of a generous welcoming of diverse views and encouragement of their free expression in an environment of civility, from an attitude that precludes any effort to sort out the truth among various prospective claims. When all views, or at least a number of conflicting ones, are accomodated and assumed to have, if not an equal, then at least some contribution to make to the truth (or call it what you

---

[15] James Collins, *Three Paths in Philosophy* (Chicago: Henry Regnery, 1962). Collins devotes the major part of three chapters (10, 11, 12) of this book to a rather well-balanced consideration of some of the questions concerning us here.

[16] Unfortunately, what frequently happens is that what may once have been a community of fellow truthseekers has now become so "balkanized" as to preclude altogether any significant conversation on important philosophic issues. Even though civility may still prevail, it is achieved largely by restricting any sharing of ideas to those other interests that people may happen to have in common.

will), it seems to me that we have abandoned all meaningful philosophic inquiry.

It should by now have become increasingly clear that this attitude of philosophical pluralism-in-principle is clearly at odds with such traditional doctrines as direct realism and the conformity theory of truth. Not only do I here acknowledge (perhaps unnecessarily) that these are the views of human knowledge and of truth to which I subscribe, but I contend that they are also the implicit convictions in accordance with which all of us live our day-to-day lives, both in private and socially, regardless of what epistemological theories we may happen to profess in an academic setting. Yet, the lack of confidence, on the part of many who had been schooled in the realist tradition of St. Thomas Aquinas, that direct realism can be upheld in the face of non-realist challenges is, I am convinced, one of their chief reasons for the widespread abandonment of Thomistic philosophy in the last half-century. A realist need not, in my judgment, attempt to defend his position by denying what seem to be obvious differences in what each of us brings to any particular encounter with the external world, nor need he hesitate to acknowledge that unique circumstances surround each such encounter. As distinct individuals, though one in nature, we are unequally endowed with human cognitive faculties. Moreover, we have in the course of our lives accumulated our own particular sets of experiences, and have come under the special influence of those with whom we share a common physical and spiritual environment. We are certainly now far from being *tabulae rasae*.

It should not, then, be surprising that, even when we encounter the same things, we see and understand them differently. Our experiences of the same things are not, and cannot be identical. It can even be argued that this is all to the good. How often have we had valuable insights to share with one another simply because we had brought with us a different perspective, and thus had formed quite different notions upon encountering, no doubt also under different circumstances, the same people, places or events?

Some may see in these observations a conflict with that often-repeated passage in the *prima pars* of St. Thomas' *Summa Theologiae* in which he likens the human intellect to prime matter in relation to the forms we receive from the things we know. Such a relationship, I grant, does prevail between the human intellect and human knowledge in general, inasmuch as I do not contribute anything of myself, anything "*a priori*," to my knowledge. This doctrine seems to me quite compatible, however, with what we have observed concerning the factors that affect the attainment of human knowledge in concrete circumstances.)

Allow me to take a moment now to come more directly to the point of these rambling observations. A realist account of human knowledge able to withstand the barrage of objections from its philosophically respectable critics must be kept focussed upon essentials. The point at issue between realism and non-realism can be stated simply as a question of what the object is that we directly encounter in our cognition. The realist's reply to that question is that what we encounter are things belonging to a domain external to our own consciousness, and that this is evident within the human cognitive experience itself. Moreover, because this is a matter of evidence, the realist will insist that there is no room here for argument,[17] beyond merely directing a would-be non-realist to his own experience.

---

[17] Not only should it be apparent that "there is no room here for argument, but I would contend, further, that neither is there any "room" for compromise with any form of non-realist thought, including (of particular interest to us here) Kantianism. I make a special point of this because the special forms of "Thomism" that Father McCool is still apparently ready to accommodate are the "transcendental" varieties (e.g. that of Bernard Lonergan) which attempt just such a compromise with the transcendental method of Immanuel Kant. See his remark on page 161: "Marechal's *Le*

The realist can do more, however. He can reject as irrelevant all arguments based on the contention that one's images or concepts always seem to fall short of being (formally) identical representations of the things we claim to know. For it is on just such grounds that many a non-realist will contend that the objects of our varied perceptions and concepts must be other than what he regards as things of an external world. But nothing of the sort really follows from the facts alleged, unless we assume (unjustifiably) that what we perceive and understand is a sort of "original" of which our perceptions and concepts are identical copies.[18]

Perhaps this last point deserves further emphasis. Once more, what is at issue in the realist vs. non-realist controversy is not the *quality* of one's cognitive representation (whether we ever have "exact" images or concepts). On the contrary, (dare I say it again) it is simply a question of what sort of object we encounter in cognition. Is it something external to my own consciousness? Is that *what* I see and understand, whether well or poorly, whether in dusk or daylight, (whether I happen to be wearing my glasses or am without them)? In other words, it is *not* part of the definition of a realist that he be a subscriber to the "faithful likeness" or "copy" theory of knowledge, a view which, frankly, appears to me utterly untenable.[19]

Point de départ de la métaphysique is shaped from beginning to end by its author's desire to begin an open and rigorous dialogue between Kant and St. Thomas.

[18] For a more extensive treatment of this matter, see my article, "What Is It To Know" in *Knowledge and Methodology*, ed. Edward D. Simmons (Milwaukee: Ken Cook Co., 1965). See also my article, "Restoring Philosophical Realism in Today's Intellectual World" in *Proceedings of the American Catholic Philosophical Association*, 1983.

[19] This is one of the few points on which I find myself in agreement with Richard Rorty. See his *Philosophy and the Mirror of Nature* (Princeton University Press, 1979).

Barring such an epistemological blunder, common human experience does not permit one even to question the universal conviction that, on both the sensory and intellectual level, we ordinarily deal directly with a world of things external to our own consciousness. That, I contend, is precisely what the realist vs. non-realist controversy is all about.

There is no denying, of course, that sometimes we fall short of achieving our normal objective of "seeing things as they are." All of us are prepared to admit that, under unfavorable circumstances (poor lighting, excessive distance, inadequate vantage point, etc.) our images of the things we see will be distorted or at least indistinct; and we are, consequently, sometimes misled into making false judgments concerning them. Error of various sorts is, after all, a fact we have learned to live with as a common hazard of human intellectual life. Most of us are quite aware that there is often a danger that we allow our preconceptions, our preferences, our prejudices to color our experience, to obscure the genuine reality. That is a failing we must learn to guard against, an obvious consequence of our finite and fallen nature, and an inevitable concomitant of the fact that our cognitive life is cumulative. But here again, the very possibility of erroneous human judgments only further attests to the fact that it is a world external to the consciousness of us all that we are dealing with even in such inauspicious circumstances. Were it not so, we would, of course, lack the grounds on the basis of which we are able to discover that we have, on a given occasion, been mistaken.

Nor ought we overlook the fact that what we carry over from past cognitive achievement, whether by direct discovery or as acquired through learning indirectly from the discoveries of others, greatly enhances our present capacity to acquire knowledge. It is difficult to imagine how paltry any given present experience of our own would be if we had to come to it untutored by our own prior knowledge, and by the rich cognitive endowment we have inherited from the wisdom of the ages.

Those are quite correct, then, who, rejecting a strictly individualist account of the quest for true wisdom, point out that communitarion aspect of human cognitive life. It would be vain--and foolhardy--to embark on a solitary search for wisdom. To argue, however, that we must be satisfied with a plurality of philosophical systems in opposition to one another, as the best possible human cognitive attainment, on the grounds that all human systems of thought, and the judgments that comprise them must forever be deficient, strikes me as a strange doctrine indeed. Apart from the obvious observation that the theory seems clearly doomed sooner or later to self-destruct in the vain attempt to accommodate incompatible views, it seems clear that anyone who holds it has abandoned the quest for truth as beyond human competence. Instead we are urged to accept wholeheartedly the present philosophical world of often irreconcilable philosophies as the ultimate attainable goal of the human search for wisdom.

From the perspective of an older philosophical tradition, this is clearly a counsel of despair. Even so, that is hardly sufficient grounds to convince someone who honestly thinks that he is, nonetheless, simply assessing human cognitive potentiality correctly. Nor would it suffice for us to point out once more that our pluralist adversaries are obviously alleging the *truth* of their theory, thus falling back upon *our* conception of truth, and of the human capacity to attain it, as the only possible means of enunciating their position. The argument on behalf of realism is more than a mere *reductio ad absurdum*.

On the other hand, we must not be too easily satisfied that we have given our opponent his due. He may, for example, think to remind us that "sometimes two heads are better than one." And if so, why not more than two? He could ask, after all, whether we have not found on occasion that we have grasped only "part of the truth" in something seen or heard, and mistaken it for the whole truth, until someone else came along to show us what we had missed? All that is asked of us, he might say, is that we acknowledge that something like this might happen also in the attempts of philosophers to accomplish their quest for wisdom. Would it not seem, then, to be the better part of prudence for a person to be always hesitant when tempted to make any claim to have achieved philosophic truth? And if this is what the pluralist position is all about, don't we seem, for no good reason, to have gotten unduly alarmed?

Once more, let us make sure we are not being misunderstood. Surely, no claim either to exhaustive knowledge or to infallibility is implied in our having rejected the view that a plurality of philosophical systems is required to overcome an alleged inadequacy and inevitable distortion of every individual perspective. On the contrary, I contend that only because we are individually capable of arriving at the truth of things can we make any sense out of either error or the possibility of its detection and correction. For it is as individuals that we encounter things themselves in cognition, and thus are enabled to make either true or false judgments about them.

In short, there seems to be no reason for one who subscribes to direct realism, and has, with the aid of St. Thomas, come to understand this position, to adopt an apologetic attitude in the face of those less confident of the capacity of the human mind for truth. This seems particularly obvious, given the fact that this philosophy of being, of man and of human knowledge is the only one in accord with common human experience, the only one by which anyone, philosopher or not, can effectively live. And the only one, as has been observed above, according to which we all do actually live.

A similar observation seems indicated concerning the notion of truth. I have long considered it to be our particular good fortune, when explaining Aristotle's famous definition of truth to undergraduates, that we have available the words today's young people so aptly hurl at one another: "Tell it like it is!" Young or old, to all of us it is evident that this is precisely, if somewhat ungrammatically, the age-old charge to speak the truth--"to say of what is, that it is, of what is not, that it is not," as Aristotle put it. And if truth, then, requires that we conform our minds (and our speech) to the way things are, it takes but a moment's reflection to notice how this notion is bound up with the realist view of knowledge. For only if we are in touch with the things that are can we "tell it like it is."

Yet, if all this is so obvious, it seems only fair to ask: Why has realism come to be so disdained today as an archaic mode of thought, in no way suited to a scientifically advanced, twentieth-century society? Why are we realists a vanishing breed, at least among those schooled in philosophy? The fact seems the more remarkable, since the issue has been more than adequately dealt with from the standpoint of philosophy and its modern history, by such of our realist predecessors as Gilson, Maritain and Mortimer Adler among others.

To those of us who have become accustomed to such manifestations of the need to distinguish the evident *in se* from the evident *ad nos*, this remains an ongoing challenge to our rhetorical skills--a challenge to be pursued on another occasion. It seems to me sufficient for our present purposes merely to observe that these traditional doctrines of realism and truth have continuously proven their perennial philosophic value.

MUST THOMISM BECOME KANTIAN TO SURVIVE? A NEGATIVE RESPONSE

Leo Sweeney, S.J.

When beginning undergraduate studies in philosophy in 1940 and thereafter, I had to face the question of what Thomism is. When it became clear that for Thomas Aquinas actual existence has primacy in all areas of metaphysics (because it has primacy in all existents) but also that he apparently elaborated that primacy while engaged in theological meditation upon *Exodus* 3, 14, and other Scriptural texts, my twofold task was to discern why a doctrine discovered by a theologian does not belong of its very nature to theology and, second, what role Aquinas is to play in one's own philosophical pursuits. In regard to the first, the doctrine of existence as fact, evidence and actualizing component within all material beings is not a strict mystery, transcending the grasp of human reason. No, the illumination which Thomas experienced while reflecting upon *Exodus* and other texts did not superimpose any intelligible content upon the data already gained from material existents through direct experience. Rather, its illuminating function was to enable him to see what actually was already contained within that data but heretofore overlooked by previous thinkers. Second, Aquinas' role with reference to subsequent students can be conceived somewhat on a parallel--having discovered the nature and primacy of existence, he now can point out to us what to look for within the data delivered by the actual world.

That twofold discernment resulted in this realization, expressed in my 1965 *A Metaphysics of Authentic Existentialism*:

> His [Aquinas'] function with reference to philosophy is, then, just this: to indicate evidence we might otherwise miss. But we formulate and assent to conclusions only if we see that evidence and if it warrants them. In general, a Thomistic philosopher is one who finds that the evidence adduced by Thomas for his stand on existence and other matters is still genuine and valid, and who then elaborates the same conclusions. If he should accept that position simply because it is Thomas', he may be a Thomist, but he is no philosopher.[1]

In the 1990's one must again face in what Thomism consists because of Gerald A. McCool's book *From Unity to Pluralism: The Internal Evolution of Thomism*,[2] which insists that traditional Thomism has died but is reborn in the Neo-Kantian theologies of Karl Rahner and Bernard Lonergan (*ibid*., "The End of the Neo-Thomistic Movement," pp. 224-30).

My current paper, obviously, is an attempt to respond to that insistence. A negative response would be easier to set forth and defend if the issue were directly between Thomas Aquinas and Immanuel Kant himself (instead of J. Maréchal, K. Rahner and B. Lonergan, as McCool wishes). Why so? Because Immanuel Kant is directly opposed to Aquinas, since for the latter the known, not the knower, is the sole content-determining

---

[1] Leo Sweeney, S.J., *A Metaphysics of Authentic Existentialism* [hereafter: *AMAE*] (Englewood Cliffs, N.J.: Prentice-Hall, Inc., 1965), p. 75, n. 25, which culminates a development begun on p. 74.

[2] New York: Fordham University Press, 1992.

[3] Here we are speaking, of course, not of mathematics, logic and empiriological sciences (which are constructural because the mind affects the content of these knowledges) but of our spontaneous and philosophical knowledge. See *AMAE*, p. 137, n. 10: "What is constructural knowledge? From reflection upon

cause of knowledge,[3] whereas for the former the knower through his/her twelve categories and the *a priori* forms of space and time determines the intellectual content of the known, which is thus solely the "phenomena" constructed by the mind and not the "noumena," which remains unknown.[4] Consequently, Kantianism so taken seems so opposed to Thomism that "transcendental Thomism" is almost a contradiction in terms.[5] In order

---

various sorts of constructs found in logic, mathematics, physics, and other empiriological sciences, as well as in our spontaneous knowledge of physical and moral evils, constructural knowledge seems to consist in these two factors: (a) The mental activity involved helps constitute (and thus affects) the very content of the intelligibilities in question, within which it shows up as an integral part. (b) Consequently, the basis of those intelligibilities is not directly any actually existing item, but rather that mental activity. By contrast, in nonconstructural knowledge (a) the mental activity involved is merely the means the knower uses to receive the intelligible message which actual existents themselves deliver; (b) accordingly, the basis of such intelligibilities is directly those actual existents." Also see *ibid*., p. 6, n. 10; L. Sweeney, S.J., *Authentic Metaphysics in an Age of Unreality*, 2nd ed. [hereafter: *Authentic Metaphysics*] (New York/Bern: Peter Lang Publishing, Inc., 1993), p. 6, n. 11; p. 146, n. 10. For Kant all knowledge is constructural.

4 On Kant's own position see Francis P. Fiorenza, "Introduction" to Karl Rahner, *Spirit in the World*, [hereafter: Fiorenza] William Dych, trans. (New York: Herder and Herder, 1968), pp. xxxv-xxxvii; James Collins, *A History of Modern European Philosophy* (Milwaukee: Bruce Publishing Co., 1954), especially pp. 468-83. Also see notes 24-26 below.

5 For Leslie Dewart any Kantianism--even that of Karl Rahner, Bernard Lonergan and Emerich Coreth--is in opposition to Thomism--see L. Dewart, *The Future of Belief* (New York: Herder and Herder, 1969), Appendix 2: "On Transcendental Thomism," pp. 499-522. For example, *ibid*., p. 504: "Transcendental Thomism cannot lay a historically valid claim to the name of Thomism"; pp. 505-6: "The idea that St. Thomas might have remotely agreed that [Karl Rahner's] being is 'being-present-to-self' . . . or even the idea that this is implicit in the doctrine of St. Thomas, cannot lie in

to understand and establish this point, let us consider a contemporary author who in doctrine is closer to Kant himself than are Maréchal, Rahner or Lonergan and whose approach will allow me simultaneously to formulate Aquinas' own epistemological position in preparation for dealing later with Maréchal and, briefly, with the other two.

Victor Preller

In 1967 Victor Preller of the Department of Religion at Princeton University published *Divine Science and the Science of God: A Reformulation of Thomas Aquinas*,[6] the first two chapters of which are especially relevant: "A Cautionary Note and Assorted Promissory Notes" (pp. 3-34) and "Epistemology Reformed and a Cautionary Note Reissued" (pp. 35-107). In them I shall concentrate upon his interpretation of Aquinas' epistemology and psychology.

A human being for Aquinas, as Preller reads him, is like an angel (pp. 44, 49, 50, 51-56, 64). This entitative condition has, of course, repercussions upon human cognition. There is innate in each human knower

---

peaceful juxtaposition with St. Thomas's doctrine" on existence as an act; p. 507: "Rahner's doctrine of the nature of human consciousness and of its object is neither implicit nor in any other way potentially contained in the doctrine of St. Thomas." For an exposition and critique of Dewart's own position on philosophy and theology, see William J. Hill, *Knowing the Unknown God* [hereafter: *Knowing*] (New York: Philosophical Library, 1971), "Leslie Dewart: Extreme Empiricism," pp. 98-109. According to Lawrence K. Shook, *Etienne Gilson* (Toronto: Pontifical Institute of Mediaeval Studies, 1984), p. 171: Gilson "never considered cartesian, kantian, lockeian or phenomenalist [*read*: phenomenologist?] thomists to be authentic thomists."

[6] Princeton, N.J.: Princeton University Press, 1967. My discussion of Preller's position in the following paragraphs is based on my review of his book--see *Modern Schoolman,* 48 (1971), 267-73.

a conceptual system which consists of intentional forms of categories (pp. 43, 53-55, 62-63) and which as such does not directly pertain to actual things [as noumena] (pp. 38, 51, 70). That system has not arisen from things (pp. 51, 52, 54) but from the agent intellect, which "all men possess *by nature* [as] a common power of conceptualization . . . and is the same in all men" (p. 70, italics in original; also see pp. 53, 55, 76).[7] Elements within that conceptual system have only a syntactical or logical role--that is, "the 'meaning' of any concept or statement which could be formed within the system would be nothing but the syntactical role played by that concept or statement within the system" (pp. 44-45). Such a system "would be entirely formal--it would have no material content" (p. 45). But can it have no reference to man's empirical experience? Yes, through the imagination, which "is naturally programmed in such a way" as to synchronize "only such regularly repeated patterns of the state of the sensory system as are isomorphic with elements within the conceptual system of the man-angel's intellect" (p. 50).

The "sensory system" just mentioned is fitted into a human being's angelic nature only after rather radical transformation. Of themselves the senses are not powers of knowing (p. 57): their content is nonintentional and apparently consists merely of the "material alterations of the receiving organs of the sensory systems--alterations caused by the impact on the organs" of external objects (p. 49). Sense knowledge

---

[7] Preller makes Aquinas' notion of agent intellect seem much like Avicenna's *dator formarum*, which was a sublunar separate Intelligence "ceaselessly radiating all possible forms and causing them to exist in proportionate matters, or to be known by intellects" (E. Gilson, *History of Christian Philosophy in the Middle Ages* [New York: Random House, 1955], p. 214; also see pp. 204-5) Aquinas often rejects Avicenna's view--see, for example, *S.T.*, I, qu. 79, art. 4 resp.; I, qu. 84, art. 4 resp.; *De Veritate*, qu. 10, art. 6 resp.

requires that the intellect combine with the senses. Only then is there "an intentional interpretation of the objects of experience, ordered to the intelligible categories of the mind" (p. 39). In fact, "perception (*conscious* experience of external reality) arises *jointly* out of the non-intentional contents of sensation and the forms of conceptual thought" (p. 62). The intelligible categories and the conceptual forms just mentioned are not caused by sense experience, which is "the *occasion* for the actual development of a conceptual system" and nothing more (p. 54).[8]

If my previous paragraphs are to some degree adequate, anyone acquainted with Thomas' Latin text can rather easily judge that Preller's exposition is incorrect. Aquinas makes abundantly clear that a human entitatively is not an angel. The human soul, he says, specifically differs from an angel (*Summa Theologiae*, I, qu. 75, art. 7 resp.; also see parallel texts given in Leonine-Marietti and in Ottawa editions). An angel is a subsistent form, which has no connection whatsoever with matter (*ibid.*, I, qu. 50, art. 1 and 2). A human being is essentially a composite of form and matter: his soul (that which is the principle of his intellectual life, as well of course as of his sensitive and vegetal life) is by its very nature the substantial form of his body (*ibid.*, I, qu. 76, art. 1 resp.). This relation of an angelic and of a human existent to matter determines the nature of knowledge in each. Since an angel is entirely immaterial, its intellect is a cognitive power which has no corporeal organ and which is in no way joined to matter (*ibid.*, I, qu. 85, art. 1 resp.: "Quaedam autem virtus cognoscitiva est quae neque est actus organi corporalis, neque est aliquo modo corporali materiae conjuncta, sicut intellectus angelicus"). Consequently, its proper object is also whatever is entirely immaterial (*ibid.*, I, qu. 84, art. 7 resp.). But a human being is mind-in-matter, an

---

[8] For Kant's influence on Preller see n. 15 below.

incarnate intellect, the proper object of which accordingly is intelligibilities concretized and existing in matter (*ibid.*, "Intellectus autem humani, qui est coniunctus corpori, proprium obiectum est quidditas sive natura in materia corporali existens"). These concretized intelligibilities are the content-determining-cause of human knowledge by acting upon the senses so as to help produce sense knowledge and by acting through the senses so as to help produce intellection (the senses and the intellect being efficient causes of their knowledges). The senses, then, are truly powers of knowing: conjoined to corporeal organs, they know material things precisely as individual and particular (*ibid.*, I, qu. 85, art. 1 resp.: "Quaedam enim cognoscitiva virtus est actus organi corporalis, scilicet sensus. Et ideo obiectum cuiuslibet sensitivae potentiae est forma prout in materia corporali existit. Et quia huiusmodi materia est individuationis principium, ideo omnis potentia sensitivae partis est cognoscitiva particularium tantum"). A human being's intellect, although itself without an organ, is nonetheless a power of a soul which is the substantial form of matter, and, hence, its proper object (as has already been said) is intelligibilities existing in matter but (this has not yet been said) not precisely as so existing (*ibid.*: "Intellectus . . . humanus . . . non enim est actus alicuius organi, sed tamen est quaedam virtus animae, quae est forma corporis . . . . Et ideo proprium eius est cognoscere formam in materia quidem corporali individualiter existentem, non tamen prout est in tali materia"). Intellection, then, is abstractive, for abstraction is just that: knowing what actually exists as individualized but without considering its individuality. But phantasms disclose to the knower an individual existent precisely as individual. Hence, our intellect efficiently brings about knowledge of material existents only by abstracting from phantasms (*ibid.*: "Cognoscere vero id quod est in materia individuali, non prout est in tali materia, est abstrahere formam a materia individuali, quam repraesentant phantasmata. Et ideo necesse est dicere quod intellectus noster intelligit materialia abstrahendo a phantasmatibus").

As that text suggests, phantasms are crucial to Thomas' theory of cognition. But what are they? What role do they play? As we previously noted, individual material existents (e.g., this running man, this roaring lion, this fragrant red rosebush) are the proper object of human knowledge, whether sensation or intellection (*ibid*.: "Objectum cuiuslibet sensitivae potentiae est forma prout in materia corporali existit . . . . Proprium eius [intellectus humani] est cognoscere formam in materia quidem corporali individualiter existentem . . ."). Such existents are intelligibilities concretized in and individualized by matter. By our external and internal senses we know such existents precisely as concretized and particularized (*ibid*.: "Et quia huiusmodi materia est individuationis principium, ideo omnis potentia sensitivae partis est cognoscitiva particularium tantum"). That is, the content of the initial, entitative and noncognitive actuation (which arises when this running man or this roaring lion or this fragrant red rosebush physically stimulates our sense organs and which informs and actuates our sense faculties) and of the ensuing sensation is "this running [man]" or "this roaring [lion]" or "this fragrant red [rosebush]."[9] The

---

[9] The bracketed word in each of the phrases indicates what is genuinely but only implicitly contained in the initial, entitative and noncognitive actuation and in the sensation, waiting to be explicated in and by the process of intellection. Or one might say that the bracketed noun points to what is virtually present in the initial entitative actuation and in the sense experience itself and which will become formally present through intellection. But no matter whether the couplet used is implicit/explicit or virtual/formal, the important point is that the content of both sensation and intellection is radically the same ("this fragrant rosebush"), although the former centers on "this fragrant," the latter on "rosebush." In Aquinas' terminology the bracketed word designates a *sensibile per accidens* (in contrast with a *sensibile per se*, whether *proprium* or *commune*)--namely, a factor which is

content of the phantasm, which is the cognitive actuation efficiently produced by the internal senses (especially the cogitative power), is no exception. Its content too is "this running [man]" or "this roaring [lion]" or "this fragrant red [rosebush]." But by intellection we know such individual material existents precisely and explicitly as actually intelligible--that is, as "man" or "lion" or "rosebush." How does this come about? The agent intellect, serving as principal efficient cause, uses as instrumental formal cause a phantasm whose content is "this running [man]" or this roaring [lion]" or "this fragrant red [rosebush]" so as to produce in the possible intellect the initial entitative actuation whose content is "man" or "lion" or "rosebush."[10] The possible intellect, because of the formal actuation thus produced within it and now informing and actuating it, has become what it will know--a becoming which is not physical (men, lions or roses are not actually in it) but intentional.[11] Through that actuation the possible or recipient intellect is genuinely although nonphysically conformed to and assimilated with the actual thing. This conformity and assimilation issue into knowledge when the intellect

---

closely involved with a *sensibile per se* (a physical object which stimulates and determines our sense powers) and is immediately apprehended in connection with the latter. See Aquinas, *In II De Anima*, lect. 13 (Marietti ed., nos. 393-96); *In IV Sent.*, dist. 49, qu. 2, art. 2 resp. (Parma ed., vol. VII, 1202a); *S.T.*, I, qu. 17, art. 2 resp.; *ibid.*, I, qu. 78, art. 3 ad 2. See G. P. Klubertanz, "St. Thomas and the Knowledge of the Singular," *The New Scholasticism,* 26 (1962), 135-66.

[10] See *Quaestio Quodlibetalis*, VIII, qu. 2, art. 3 resp. (Marietti ed., p. 161): "Intellectus agens est principale agens, quod agit rerum similitudines in intellectu possibili. Phantasmata autem quae a rebus exterioribus accipiuntur, sunt quasi agentia instrumentalia: intellectus enim possibilis comparatur ad res quarum notitiam recipit, sicut patiens quod cooperatur agenti."

[11] This stage of the process is precognitive with reference to the intellection which has not yet occurred.

efficiently causes the concept or *verbum*, the explicit content of which is "man" or "lion" or "rosebush."[12]

In such an epistemology and psychology there manifestly are two causes of human knowledge. The cognitive faculties (external and internal senses; the agent and possible intellects) are its efficient cause, but actual individual things are its content-determining-cause.[13] A

---

[12] See *Quaestio Disputata De Veritate*, qu. 1, art. 1 resp. (Leonine ed., XXII, vol. 1, pp. 5-6): "Omnis autem cognitio perficitur per assimilationem cognoscentis ad rem cognitam, ita quod assimilatio dicta est causa cognitionis: sicut visus per hoc quod disponitur per speciem coloris, cognoscit colorem . . . . Hoc est conformitatem, ut dictum est, sequitur cognitio rei. Sic ergo entitas rei praecedit rationem veritatis, sed cognitio est quidam veritatis effectus."

[13] The distinction between efficient and content-determining-causes is based upon such texts as *De Veritate*, qu. 10, art. 6 resp. (Leonine ed., XXII, vol. 2, pp. 312-13): "[On the question of 'utrum mens humana cognitionem a sensibilibus accipiat'] rationabilior est sententia Philosophi qui ponit scientiam mentis nostrae partim ab intrinseco et partim ab extrinseco esse, non solum a rebus a materia separatis sed etiam ab ipsis sensibilibus. Cum enim mens nostra comparatur ad res sensibiles quae sunt extra animam, invenitur se habere ad eas in duplici habitudine: uno modo ut actus ad potentiam, in quantum scilicet res quae sunt extra animam sunt intelligibiles in potentia, ipsa vero mens est intelligibilis in actu, et secundum hoc ponitur in anima intellectus agens qui faciat intelligibilia in potentia esse intelligibilia in actu; alio modo ut potentia ad actum, prout scilicet in mente nostra formae rerum determinatae sunt in potentia tantum quae in rebus extra animam sunt in actu, et secundum hoc ponitur in anima nostra intellectus possibilis cuius est recipere formas a rebus sensibilibus abstractas, factas intelligibiles in actu per lumen intellectus agentis, quod quidem lumen intellectus agentis in anima procedit sicut a prima origine . . . et praecipue a Deo. Et secundum hoc verum est quod scientiam mens nostra a sensibilibus accipit; nihilominus tamen ipsa anima in se similitudines rerum format in quantum per lumen intellectus agentis efficiuntur formae a sensibilibus abstractae intelligibiles actu, ut in intellectu possibili recipi possint." For a discussion of the distinction between efficient

human knower directly and noninferentially knows actual things because they have caused the content of his knowledge and this they can do because they are concretized intelligibilities. Thus, they account for what he knows. Thomas' position on cognition, therefore, is not Kant's. He neither needs nor permits any categories or conceptual forms or first principles which in origin would be in any way independent of actual existents. A second feature of Thomas' theory is that the initial formal actuations are everywhere required as entitative determinants in the intentional order of the cognitive faculties. By them our powers entitatively, intentionally and precognitively become what will be known. Third, the phantasm has several facets. It is the awareness which is efficiently produced by man's internal senses but whose content consists of data from the external senses as well. By it one knows, to use a previous example, this fragrant red rosebush as "this fragrant red [rosebush]." Because of its content it also can serve as the instrument which the agent intellect uses to produce within the possible intellect the entitative actuation whose content is "rose." Because its content is itself determined by the red rosebush affecting the external and internal sense organs, it is the channel through which the actual rosebush is the content-determining-cause of that entitative actuation and, thereby, of the concept "rose."[14]

---

cause and content-determining-cause, see my *Authentic Metaphysics*, pp. 325-26. For an exposition of the metaphysics of Thomas' theory of abstraction see *ibid.*, pp. 338-46 and 405-6.

[14] On Preller's inaccurate interpretation of phantasm see the following statements. P. 50, n. 51: "The phantasm cannot be an 'intentional image' possessed 'by the mind.' Sensation cannot be a conscious state until the contents of sensation (physical states of the organism) are 'informed' by the intellect"; also see pp. 54-55. P. 56: "Aquinas seems to define the 'phantasm' in such a way that it must be both intelligible and nonintentional--both material and (in some sense) mental. I shall attempt to show that the very notion of a perceptual mental image . . . prior to the operation of

If the previous three paragraphs are correct, one must conclude that Preller has misread Aquinas. A human being having an *a priori* system of conceptual forms and categories within his mind; the senses not causes but only occasions of that system; the senses not themselves powers of knowing; the entitative, noncognitive initial actuation of the faculty (whether sense or intellect) not distinguished from the subsequent cognitional actuation of the same faculty--these points are in conflict with the medieval theologian's texts. And they also reveal that Preller has obviously used Kant in approaching Aquinas to such an extent that the latter is scarcely recognizable.[15]

---

the intellect is self-contradictory"; see pp. 58-59. P. 61: "The 'phantasm' is *simply* Aquinas' method of . . . indicating how the physical may affect the 'mental' . . . . I think we must [grant] that it is an impossible entity--a self-contradictory *tertium quid* . . . . The most crucial aspect of the 'logic of phantasms' is that the 'phantasm' is produced by a kind of pre-judgmental and pre-conceptual automatic causality . . . . Nevertheless, the form that the 'image' actually takes is dictated by the presence 'in the mind' of conceptual powers and categories. All that Aquinas can really be saying is that our *conscious experience* of, e.g., 'blue objects' is a result *both* of that which occurs in the physical sensory system when we sense a blue object, *and* of the production by the intellect of the formally significant concept 'blue.' *There are no 'phantasms'*" (italics in original).

[15] Kant's influence on Preller came not only through his own study of Kant but also through Wilfrid Sellars, whom he acknowledges using to "translate the epistemological doctrine of the *Summa Theologiae* [of Aquinas] into a fairly contemporary mode of discourse highly dependent on the insights and terminology of Wilfrid Sellars" (Preller, p. 36). For Kant's place in Sellars' thought, see Wilfrid Sellars, *Science, Perception and Reality* (London: Routledge and Kegan Paul, 1963), pp. 46, 73n, 74n, 90, 100-101, 127, 299. Also see *idem*, *Science and Metaphysics: Variations on Kantian Themes* (London: Routledge and Kegan Paul, 1968) in its entirety. For Sellars' own intellectual autobiography see his "Autobiographical Reflections" in Hector-Neri Castañeda (ed.), *Action, Knowledge and Reality: Critical*

Those points also prepare us, I hope, to understand and evaluate the Neo-Kantianism of Joseph Maréchal and of those he influenced.[16]

Joseph Maréchal

> Joseph Maréchal (1878-1944) perhaps more than any other Catholic thinker . . . has influenced the direction that the question of man's cognitive groping for God has taken in contemporary times, and his influence is a continuing one. At least more profoundly than others, he set himself the task of facing up to the dilemma into which the critical studies of Kant had thrown the traditional epistemology derived from Scholasticism.[17]

Studies in Honor of Wilfrid Sellars (Indianapolis: Bobbs-Merrill Co., Inc., 1975), pp. 277-93 (on Kant, see pp. 283-89 especially). Also see Johanna Seibt, *Properties as Process: A Synoptic Study of Wilfrid Sellar's Nominalism* (Acascaders, Cal.: Ridgeview Publishing Co., 1990.

[16] Prominent among the latter would be Karl Rahner (1904-1984) and Bernard Lonergan (1904-1984), the first of whom is influenced not only by Kant and Maréchal but also by Heidegger, the second also by modern science, philosophy of history and John Henry Newman. See William J. Hill, "Thomism, Transcendental," *New Catholic Encyclopedia* 16 (1964), 451-53; Fiorenza, pp. xix-xlv; William J. Hill, *Knowing*, pp. 66-79 (on Rahner), pp. 79-88 (on Lonergan); McCool, pp. 224-30. On the intellectual development of Maréchal see André Hayen, "Le Père Joseph Maréchal (1878-1944)," in *Mélanges Joseph Maréchal* (Bruxelles: L'Édition Universelle; Paris: Desclée de Brouwer, 1950), pp. 3-21; A. Milet, "Les premiers écrits philosophiques du P. Maréchal (1901-1913)," *ibid.*, pp. 23-46. On K. Rahner see Donald L. Gelph, *Life and Light: A Guide to the Theology of Karl Rahner* (New York: Sheed and Ward, 1966), ch. 1, pp. 3-14; on B. Lonergan see his own account, "Insight Revisited," in *A Second Collection by B. J. F. Lonergan*, ed. W. J. Ryan and B. J. Tyrell (Philadelphia: Westminster Press, 1974), pp. 263-78.

[17] Hill, *Knowing*, p. 59. This book proves to be clear, accurate and amply documented. McCool nowhere refers to it.

This Belgian Jesuit realized that "the Kantian *Critiques* . . . had to be taken seriously and it is against this precise background that Maréchal attempted to reintroduce the whole epistemological problem" in especially his *Le Thomisme devant la philosophie critique* (Louvain/Paris: Museum Lessianum/Alcan, 1926; 2nd ed., 1949). What this volume indicates is "less an historical interpretation of St. Thomas than a reinterpretation of the basic Thomistic insights in the light of the Kantian crisis."[18]

[1] In that reinterpretation human intellection remains dependent upon sensation.[19] But the nature of the intellect

> is such that it has a natural, spontaneous agency (*intellectus agens*) in the face of sensible receptivity, which issues inwardly in an abstract essence. This latter is a pure *representation*, immanent to the knower, so that from it there cannot be inferred the extramental reality or existence of what is represented. The representation, then, is totally phenomenal in kind, and possesses of itself no value of the real. Already there is the suggestion that what man is first of all cognitively aware of is his own concepts, not realities themselves.

[2] But how can the intellect break out of this mold of subjectivity? Only "by recourse to an activity other than this primal one of 'representing' in mental constructs. This is *affirmation*, wherein the human spirit 'drives through' the concept, as it were, to make contact

---

[18] *Ibid.*, p. 60. Arabic numerals in brackets are inserted in the next three paragraphs for ease in reference.

[19] McCool, p. 87, corroborates that fact: "Maréchal shares Rousselot's Thomistic metaphysics of the phantasm, agent intellect, impressed *species*, and the mental word of the concept and judgment."

with the noumenal order."[20] [3] This activity is not that of the human will but is explicitly identified "as an activity of the intellect, consisting actually in the judgment," seen however not as a "mere synthetic activity [of formulating propositions] but [of] objectifying: something midway between an exclusively synthetic activity reaching only to the formal aspect of being and an intellectual intuition knowing *being* in full." That is, the full role of the judgment "is to achieve an objective synthesis that is dynamic and trans-categorical. [4] The *representations* immanent to our thought have no value of object apart from an implicit *affirmation*, which, in the largest sense of the word, is the active reference of a conceptual content to reality." Thus the intellect achieves "in affirmation a virtual prolongation of its own movement beyond the frontiers of the subject [and] to the thing.[21]

[5] For Maréchal the affirmation implicitly presents the opposition of subject to object "and thus constitutes the object as object in consciousness and attaches it to the ontological order." [6] Thus the knower comes to discern "the innate tendency of his intelligence through and beyond the contents of his concept to a trans-subjective order of being outside the knower." And this tendency "is the indigenous drive or appetition of spirit towards termination at real being beyond the phenomenal being of consciousness that founds the affirming judgment." [7] Accordingly, the dynamism of the "cognitive faculty towards its own proper object . . . [is] radicated in the natural antecedent finality of the integral subject possessing the noetic power."[22]

---

[20] Hill, *Knowing*, p. 60.

[21] *Ibid.*, pp. 60-61.

[22] *Ibid.*, p. 61. But as Hill accurately points out, Maréchal's affirmation as a dynamic and trans-categorical bridge between subject and extramental reality appeals "to an activity within the knower that is itself non-cognitive" (p. 62) and thus "is not a cognitive act at all . . . but rather a 'projective' act or a

[8] And this "projection of the contents of an intentional representation onto the entitative order demands at the same time the affirmation of absolute and infinite Being." Thus the tendency of the intelligence is "towards the Absolute . . . and founds all other 'affirmations' or judgments of reality." In this way the human intelligence "is able to reconstruct a noumenal unity that is the analogical and universal unity of being, which in its turn is founded on the Absolute Unity." [9] Consequently, the intellect's affirmation of finite being actually posits "existents in the extra-mental real world at some relative point on the ontological scale with a necessary reference of greater or lesser remotion (in excellence) from Absolute Being. Thus the judgment of reality preserves an analogical character, one wherein the prime analogue is always present and measuring the being affirmed of lesser realities."[23]

Thus runs Maréchal's position, which allows itself to be summarized under these three main points.

'positing' by the subject--somewhat akin to Kant's postulates of the practical reason" (p. 63). Although it is not "an activity of the will . . ., [the tendency within an affirmation] is the internal *natural* appetite of the rational being." Hence, there has occurred in Maréchal's theory a "subtle transposition from the cognitive order to the appetitive or conative." Thus this tendency "is indeed *of* the intelligence but it is not a 'knowing' activity . . . . Through 'desiring' the real order, man's intelligence is led to 'affirm' it without however 'knowing' it." Hence, Maréchal's intellectual dynamism is open to the charge that it is 'in fact non-intellectual and non-cognitive" (*ibid.*). On all operative powers of a human person, in Aquinas' position, as determinate and natural tendencies each to its proper object, see my *Authentic Metaphysics*, pp. 290-292. On the human will as itself such a determinate and natural tendency to the good-as-known, plus the will's operation of desiring or loving as an additional tendency, see *ibid.*, pp. 292-94.

[23] Hill, *Knowing*, p. 62.

(a) The human intellect first knows phenomenal being by forming from sensibility a concept or representation of abstract essences (#1).[24]

(b) The intellect then formulates judgments, the first function of which is "synthetic": the intellect affirms a predicate of a subject ("S = P") (#3). Within that synthetic affirmation the human knower is simultaneously conscious of his/her contrast as subject *vs.* object (thus, affirmation is an objectifying and trans-categorical process) and thereby is aware of noumenal and extramental being as its object (#2-#5).[25]

(c) But within that intellectual affirmation there resides a constant and inevitable tendency, drive, appetite (call it what you will) also to absolute and infinite Being, which the human intellect or spirit is, at least implicitly, affirming so as to measure the greater or lesser excellence of noumenal beings (#6-#9).[26]

If those three points accurately summarize Maréchal's theory, he obviously has departed rather radically from Aquinas' own epistemology and metaphysics. Why so? For Aquinas what the human person straightaway and directly knows are noumena, because as actually existing and concretized

---

[24] Here, obviously, Maréchal agrees with Kant's view in the *Critique of Pure Reason* (Norman Kemp Smith translation, pp. 257-75, which summarize pp. 65-256) that the human knower through the *a priori* forms of space and time and the twelve categories fashions from sense data, and thus knows, phenomena.

[25] Here Maréchal disagrees with Kant, for whom even synthetic *a priori* judgments concern not noumena but phenomena. See *Critique of Pure Reason*, Smith trans., pp. 48-62 and 191-97.

[26] On the kinship of Maréchal's affirmation of Absolute Being with Kant's assent in the *Critique of Practical Reason* to God's existence as a postulate of practical reason, see n. 22 above. Also see Kant, *Critique of Practical Reason*, Lewis White Beck trans., pp. 128-36; also *Critique of Pure Reason*, Smith trans., pp. 322-26 and 485-95.

intelligibilities[27] these are the content-determining-causes of human knowledge in this way:[28] they produce in each

---

[27] Each such noumenon is a single but composite entity fashioned of forms or acts actualized by existence and of matter and potencies, which concretize and individualize them and which also are positive and real factors in and of the essence. Accordingly, the forms and acts of the entity are intelligibilities illumining the human knower--see Thomas, *In Librum de Causis*, Prop. 6 (Saffrey trans., p. 45, lines 12-15): "Unumquodquod cognoscitur per id quod est in actu; et ideo ipsa actualitas rei est quoddam lumen ipsius et, quia effectus habet quod sit in actu per suam causam, inde est quod illuminatur et cognoscitur per suam causam," which is the entity or noumenon itself.

Thus, such entities or noumena are rich and not impoverished sources of data because Aquinas (and Aristotle) does not suffer from what Iredell Jenkins fifty years ago (*Journal of Philosophy*, 39 [1942], 533-47) called "the postulate of an impoverished reality": namely, "the settled conviction that nature is *in fact* much simpler and barer than it appears to us *in experience*." Accordingly, "man adds on to reality in the act of experiencing it. It thus states that the real is poorer than our experience of it; that our acquaintance with the externally real transforms and enriches the actual contents of this real; that we . . . bestow upon reality most of the richness that we pretend to discover in it. In its first, simplest and classic form this postulate is the doctrine of primary and secondary qualities, as enunciated by Descartes, Bacon and Locke. As such, it states that real things have in themselves only the properties of extension, mass, figure and motion."

Apparently and in contrast to Jenkins, Robert O. Johann ("Subjectivity," *Review of Metaphysics,* 12 [1958], 204-5) accepts this impoverished view of material things when he contrasts "being as subject" with "being as object." What is the nature of the latter? Such a being-as-thing "exists, to be sure; it is there. But its existence has no depth, is absorbed, as it were, in the pattern it actualizes, in the face it presents to the world. There is no disproportion between an inner reality and its outer manifestations, between a secret center and source of activity and any of its particular realizations. On the contrary, the impersonal is all surface, and hence accessible to all comers." Also W. Norris Clarke apparently views existents other than one's "self" as

of our operative powers (the external and internal senses, the recipient intellect) an entitative, nonphysical, formal determinant or actuation (let us call it "actuation *a*"), which makes the operative power entitatively *be* the known.[29] Thus actuated, the operative powers

---

impoverished--see "The Self as Source of Meaning in Metaphysics," *ibid.*, 21 (1968), 597-614, especially pp. 600-601.

Jean-Paul Sartre impoverishes selves and nonselves by reducing both to series of appearances without any interiority at all: "By reducing the existence to the series of appearances which manifest it . . . we certainly thus get rid of that dualism which in the existent opposes interior to exterior. There is no longer an exterior for the existent if one means by that a superficial covering which hides from sight the true nature of the object . . . . And this nature . . . no longer exists . . . . For the being of an existent is exactly what it appears" (*Being and Nothingness*, quoted in my *Authentic Metaphysics*, p. 112).

Aquinas would eschew such views: each individual material existent is a rich composite of perfections--individuated by matter, to be sure, and thus requiring abstraction to be known intellectually, but illumining both sense and intellectual faculties as their content-determining-causes. On abstraction see prgr. immediately prior to n. 9 above.

[28] See prgrs. corresponding to notes 9-14 above, which cite relevant texts from Thomas.

[29] For Aquinas to know is, then, to become and be the known intentionally, and the actuation "a" of the faculty results in the knower being united in an entitative and noncognitive fashion with the object causing that determinant prior to the cognition itself. Aquinas bases this aspect of his epistemology upon Aristotle, *On the Soul*, III, c. 2, 425b26 and 426a10: "the actuation [not "activity," as others have translated *energeia*] of what is sensed and of the sense is one and the same (although the being of each is not the same), but that actuation is in the sense and not in what is sensed." By transferring that insight from sense perception to any spontaneous or nonconstructural knowledge, we obtain this: the actuation of what is known and of the faculty knowing is one and the same but it is present in the faculty and not in the known. It is an actuation *of* the known and is caused *by* the known, while simultaneously it is also an actuation and

efficiently cause knowledge itself (awarenesses, judgments, subsequent reflections when needed to formulate additional judgments--all these are cognitive actuations, which can be called "actuations *b*").[30] The human knower then becomes aware of him/herself only in and through knowing those actual material existents.[31]

But how and when does the human person know "being"? As implied within each and every affirmation of the human spirit (as Maréchal would have it--see above, #6-#9) as the absolute Being, against which all finite beings are measured? No, for Aquinas the human person knows "being" by knowing the actually existing material beings which are the content-determining-causes

---

determination *of* the knower too, and, in fact, is solely *in* the knower. The very same actuation which pre-and non-cognitively conditions, determines, forms the faculty by being present in it is also the actuation of that which is known (precisely as known, though, and not as it is outside the cognitive process, where its actuations are its substantial and accidental forms and its act of existing) and which is causing it in the knower. Through this actuation or determination, which is *in* and *of* the knower and *of* and *from* the known, the object known is the content-determining-cause of my knowledge. See my *Divine Infinity in Greek and Medieval Thought* (New York/Bern: Peter Lang Publishing, Inc., 1992), pp. 560-61; my *Christian Philosophy: Nature, Origin, Developments* (New York/Bern: Peter Lang Publishing, Inc., 1994), ch. 22: "Preller and Aquinas: Second Thoughts," the sections on "Aristotle's Theory of Knowledge" and on "Aquinas: Existence and Actuation 'A.'"

[30] On the threefold division of cognitive operations see my *Authentic Metaphysics*, pp. 326-33.

[31] See *S.T.*, I, 87, 1 resp. and ad 3; *ibid.*, a. 3, resp.: "Est autem alius intellectus, scilicet humanus, qui nec est suum intelligere, nec sui intelligere est obiectum primum ipsa eius essentia, sed aliquid extrinsecum, scilicet natura materialis rei. Et ideo id quod primo cognoscitur ab intellectu humano est huiusmodi obiectum; et secundario cognoscitur ipse actus quo cognoscitur obiectum; et per actum cognoscitur ipse intellectus, cuius est perfectio ipsum intelligere."

of that knowledge, since he or she knows them not only as this or that sort of existent but precisely *as existent*, where actual existence is not only a fact but also an evidence of an actualizing component which is other than the essence or nature actualized, over which it has primacy.[32] Furthermore, this otherness of existence from essence indicates that such actual existents are contingent: they exist in such a way that none exists of its very nature and yet each *does* exist and is made real by that existence. Accordingly, they point to the subsistent absolute and infinite Being who is God and who creates and conserves them because the actuation of existence in each of them is His proper effect, since His very nature or essence *is* existence.[33]

---

[32] On this primacy see my *Authentic Metaphysics*, ch. 5: "Primacy of Existence in Existents"; also my *Christian Philosophy*, ch. 21: "The Mystery of Existence."

[33] See *Authentic Metaphysics*, "Actual Existence and God," pp. 135-39. As efficient (both creative and conservative) cause of existence as His proper effect, God transcends all creatures (because He is other than them as their efficient cause) and yet is entirely immanent to them--see *S.T.*, I, 8, 1 resp.: a brief but brilliant *tour de force*, in which the act of existence is said to be "illud quod est magis intimum cuilibet et quod profundius omnibus inest . . . . Unde oportet quod Deus sit in omnibus rebus et intime."

[34] For Thomas our knowing the divine Existent through material existents affects not only our philosophical but also our theological knowledge of God since divine revelations concerning God are expressed in sacred scriptures by inspired writers using human words (both interior and exterior), which are then interpreted by theologians who must rely on their human words (both interior and exterior) to express their exegesis. Little wonder, then, that Aquinas states that we can know that God exists but not what He is (*S.T.*, I, 3, Introduction). In fact, according to Thomas when we speak of God as "existence," what we directly know is solely what the verb expresses in the affirmative proposition "God exists" (see *ibid.*, I, 3, 4 ad 3; *De Potentia*, 7, 2 ad 1). All else--divine existence as subsistent, as goodness, as intellection, wisdom, volition, love and so on--is known mediately (i.e., through

Hence for Aquinas and unlike Maréchal our human affirmations do not include (even implicitly) an initial affirmation of the existence of divine Being: affirmation of God comes toward the end of philosophy when we *prove* that He exists from actually existing material beings (see above) or from Faith, when we assent to God's revealing Himself to us through the Scriptures or otherwise.[34]

"But," someone may object, "Does not Aquinas himself state that we initially have an implicit knowledge of God? For instance, does he not say that 'knowledge of God's existence is naturally implanted in us'?"[35] Granted that such is part of his statement but, more fully, it is as follows: "To know that God exists in a general and opaque way is naturally implanted in us inasmuch as God is happiness for human beings, who naturally desire to be happy, and that which is naturally desired by them is naturally known by them."[36] But in what sense is that which we naturally desire naturally known by us? Or, more exactly, how do we know that happiness consists in our achieving a good which is perfect ("perfectum hominis bonum, quod est beatitudo")? For Thomas that knowledge comes not from innate ideas of any sort because he takes seriously Aristotle's view that the human intellect of itself is a *tabula rasa*, and thus is

---

creatures, who are good, intellective, free, loving) and negatively and eminently (e.g., God is good but not merely as we are good, God is wise but not merely as we are wise, etc.). See my *Christian Philosophy*, ch. 19: "Metaphysics and God: Plotinus and Aquinas."

[35] *S.T.*, I, 2, 1 ad 1: "Cognoscere Deum esse . . . est nobis naturaliter insertum."

[36] *Ibid.*: "Cognoscere Deum esse in aliquo communi, sub quadam confusione, est nobis naturaliter insertum, inquantum scilicet Deus est hominis beatitudo: homo enim naturaliter desiderat beratitudinem, et quod naturaliter desideratur ab homine, naturaliter cognoscitur ab eodem."

dependent for its entire content upon the concrete cases in which we and other actual existents find ourselves (see note 29 above). Thus knowledge of "happiness" comes from our experiencing actual concrete cases, which put across to us what "good," "human," "knowledge," "desire," "perfect" and so forth are. Such cognition is "natural" since it has arisen from "nature"--i.e., from what we and other actual material existents are and experience. To be aware of such objects, Thomas asserts,

> is not, simply speaking, to know that God exists--any more than to be aware that someone is approaching is to be aware of Peter, even though it may be Peter who is approaching. So too to know that happiness is a human being's perfect good is not the same as knowing that God is that good since many consider it to be wealth or pleasure or some such thing.[37]

Accordingly, such a text shows Thomas' difference from Maréchal, for whom every human judgment is simultaneously an (at least implicit) affirmation of the divine and absolute Being as paradigm of finite beings (see above, #8-#9). But for Aquinas even our naturally knowing and desiring happiness is not to know God as God, who nonetheless will eventually

---

[37] *Ibid.*: "Sed hoc non est simpliciter cognoscere Deum esse; sicut cognoscere venientem, non est cognoscere Petrum, quamvis sit Petrus veniens: multi enim perfectum hominis bonum, quod est beatitudo, existimant divitias, quidam vero voluptates, quidam autem aliquid aliud." Also see *ibid.*, I, 85, 3 resp.; *De Veritate*, 10, 11 ad 10; *ibid.*, 10, 12 resp. and ad 3. See John F. X. Knasas' accurate exegesis of these other texts: "Transcendental Thomism and the Thomistic Texts," *Thomist* 54 (1990), 81-95; *idem*, *The Preface to Thomistic Metaphysics* (New York/Bern: Peter Lang Publishing, Inc., 1990), ch. 3: "Transcendental Method," pp. 47-69.

turn out to constitute our happiness. The two authors show themselves again to be quite different.

Epilogue

Our task of answering whether an epistemology and metaphysics immediately based upon Aquinas' own texts[38] can be resuscitated by Kantianism would be much easier if Maréchal and those he influenced were as patently Kantian as is Victor Preller. Unfortunately (or fortunately?), they are not. To restrict oneself to Maréchal: his position mixes Kant with insights borrowed from Fichte, Maurice Blondel, Pierre Rousselot and others.[39] The result is that he radically differs from Thomas Aquinas to such an extent that to inject any sort of Maréchalian theory (even that of Rahner or Lonergan) into him would hinder rather than help. Besides, Aquinas authentically interpreted on his own terms is, I submit, alive and well--at least, west of the Hudson.

---

[38] Notice that here I do not speak of or defend Thomism as such, because the term has had (as McCool has shown competently) a long and often unfortunate history: often what it expresses should have succumbed. But not so with what issues directly from a study of Thomas himself.

[39] See n. 16 above. On Maréchal's relationship with Fichte see Fiorenza, p. xxxiii: Gustav Siewerth "correctly noted that the Maréchalian school is closer to Fichte than to Kant . . . [Yet it is also true that] Maréchal explicitly rejects what he considers to be the 'absolute idealism' of Fichte." Also see Maréchal's posthumously published *Le système idéaliste chez Kant et les postkantiens* (Bruxelles: L'Édition Universelle; Paris: Desclée de Brouwer, 1947) [Cahier IV of *Le point de départ de la métaphysique*], ch. II: "Idéalisme transdendental de Fichte," pp. 335-455. As its editors point out (pp. 9-10), the contents of that long chapter came from Maréchal's early revisions (1917, 1918) of *Le point de départ* and hence long before its publication in 1947.

PHILOSOPHICAL PLURALISM AND "THE INTERNAL EVOLUTION OF THOMISM": SOME REALIST ANIMADVERSIONS

Denis J. M. Bradley

I. Philosophical Pluralism; *De Facto* And *De Jure*

The distinction between *de facto* and *de jure* philosophical pluralism figures, incidentally and somewhat insouciantly, in Gerald McCool's benign historical survey of "the internal evolution" of twentieth-century Thomism.[1] In fact, this distinction, but only if it is advertently, sharply, and fully drawn, is required if we are to comprehend the two theses that underlie McCool's book. The first of these theses is that "the legitimacy of pluralism in philosophy and theology" can be defended "on the basis of St. Thomas' own epistemology and metaphysics" (p. 2). Throughout his book, McCool, it should be duly noted, never speaks about his own position *in propria persona*; he attributes this first thesis only to certain of the Thomists whom he considers (Rousselot, Maréchal, Bruno de Solages, and Le Blond). But nothing in his exegesis of these Thomists suggests that he himself would repudiate their claim which, indeed, seems to frame his own history of Neo-Thomism.[2]

---

[1] See Gerald A. McCool, S.J., *From Unity to Pluralism: The Internal Evolution of Thomism* (New York: Fordham University Press, 1989), p. 11: "Problems of historical development, hermeneutics, or diverse conceptual frameworks did not trouble them," i.e., did not trouble nineteenth-century Neo-Scholastics.

[2]Cf. McCool: "Le Blond was arguing that the epistemology and metaphysics of the Angelic Doctor require as a matter of principle that there be a plurality of speculative systems in

There are, however, as many problems and puzzles latent in this first thesis as there are problems and puzzles latent in the undifferentiated notion of "philosophical pluralism." Although McCool does not even allude to them, these problems and puzzles, which are the meat and gristle of several recent symposia,[3] can hardly be ignored, at least in their broad outlines, if we wish to assess McCool's notion of the *internal evolution* of "Thomism."

That there have been, are, and, in all likelihood, will continue to be many diverse philosophies, whose peculiar doctrines are in many instances logically incompatible, is the conspicuous fact to which the term "philosophical pluralism" refers. While such contradictions enliven and sharpen philosophical discussion with past and present-day controversy, they fragment the alleged unity of reason and give the lie to its putative universality. Moreover, philosophical controversies, when enacted, ideologically divide persons and groups, strain civility, and injure the commonweal. Thus what started as invigorating academic debate can end in invidious power politics.

Contemporary interest in *de facto* philosophical pluralism, although the latter has been a philosophical commonplace since antiquity, grows out of renewed interest in the theoretical and practical significance of contemporary cultural, religious, political, and moral pluralism. Nonetheless, *de facto* philosophical pluralism

---

Catholic philosophy and theology. If Le Blond was right, the *magna charta* of the Neo-Thomistic movement could not be reconciled with the exigencies of Thomas' own thought" (p. 219); "On the basis of St. Thomas's own fundamental principles the Catholic theologian now realized that the dream of the Neo-Thomist movement could never be realized" (p. 228).

[3] See, for example, *Rationality and Relativism*, ed. Martin Hollis and Steven Lukes (Cambridge, Massachusetts: MIT Press, 1982); *Is Relativism Defensible?*, *Monist*, 67, no. 3 (1984); *Systematic Pluralism*, *Monist*, 73, no. 3 (1990).

poses no threat to the putative universality and coherence of philosophical reason if it is maintained, from any one philosophical standpoint, that opposing doctrines can be critically or "dialectically" elaborated and thus exposed as so many errors (Aristotle) or, in a comprehensive dialectic, subsumed and reconciled as so many partial truths (Hegel).[4] Some use of "dialectical negation" has been part of the repertoire of most philosophers major and minor who, like Plato, Aristotle, Kant, and Hegel, have not hesitated to dispatch rival philosophies in the name of *the* truth. Yet, radical or *de jure* philosophical pluralists, of whom the late Richard McKeon was certainly the most erudite and, notoriously, the most subtle, are full of principled hesitation--like the ancient sophists, skeptics, and rhetoricians--about embracing any exclusionary notion of philosophical truth.[5] Such "truths" have proved to be only too dubious; the history of philosophy is the cycle of philosophical refutations which are in turn refuted.[6] But these refutations, upon

---

[4] Cf. Aristotle, *Topics*, I, 2, 101b2-4; Hegel, Introduction to *Logic (Encyclopedia of the Philosophical Science).*

[5] See Richard McKeon, "Dialogue and Controversy in Philosophy," *Philosophy and Phenomenological Research*, 17 no. 2 (1956), 143-163; reprinted in *Freedom and History and Other Essays: An Introduction to the Thought of Richard McKeon*, ed. Zahava K. McKeon (Chicago and London: The University of Chicago Press, 1990), pp. 103-125: "The primary purpose of philosophy is the discovery and demonstration of truth. But 'truth' is differently conceived according to the principles of different philosophies, and philosophical methods are constructed to form and justify bodies of doctrines which express truths so conceived and so justified" (p. 122).

[6] Cf. Gottfried Martin, *General Metapysics: Its Problems and Method*, trans. Daniel O'Connor (London: George Allen & Unwin Ltd, 1968), pp. 326-327: "The history of philosophy testifies unmistakably for the plurality [of metaphysical standpoints]. Whatever be the significance of Plato, or Aristotle, and of Kant, after the subsequent course of philosophy it is not to be expected that one of these three great standpoints will be

examination, often do not actually touch the philosophies being refuted; they typically proceed by first restating, using diverse principles and method, opposed doctrines which when restated, in an alien philosophical medium, appear to be blatantly contradictory or meaningless.[7] McKeon cautions that dialogue among philosophers, lest it degenerate into otiose controversy, cannot assume the peculiar categories, or method, or doctrines of any one philosophy; it can only start by interpreting what is common, i.e., what a philosopher *says* about things, thought, actions, and language.[8] But interpreting what is said requires the clarification of the common but ambiguous terms that are used by every philosopher to specify and to explain his or her peculiar problems with the putative facts about things, thought, actions, and language.[9] Clarification of philosophical speech occurs, McKeon claims, when we specify the "semantic schemata" that determine the "subject-matter and problems of [a] philosophy."[10]

De facto philosophical pluralism, if it cannot be dialectically resolved, poses a radical theoretical problem, or, in Aristotle's sense, becomes aporematic,[11]

---

recognized as the only correct one and that this will permit the other two standpoints to be proved false."

[7] See Richard McKeon, "Philosophy and Method," *The Journal of Philosophy*, 48, no. 22 (1951), 653-682; esp. 659.

[8] See Richard McKeon, "Discourse, Demonstration, Verification, and Justification," in *Demonstration, vérification, justification*, Entretiens de l'Institut International de Philosophie, Liège, Septembre 1967 (Louvain: Editions Nauwelaerts, 1968), pp. 37-63; reprinted in *Rhetoric: Essays in Invention and Discovery*,

[9] See Richard McKeon, "Philosophy of communications and the Arts," in *Perspectives in Education, Religion, and the Arts*, ed. H. Keifer and M. Munitz (Albany: State University of New York, 1970), pp. 329-350; reprinted in *Rhetoric*, pp. 95-120.

[10] *Ibid.*, p. 105.

[11] For a discussion of the Aristotelian term *'aporia'*, whose primary meaning is "lack of (intellectual) passage" (*Met.*, II, 1,

to any philosopher who assumes the unconditional universality and coherence of philosophical reason. Among the welter of competing doctrines, where does reason find a "clear passage" to *the* truth? The assertion of radical or *de jure* philosophical pluralism overturns this "dogmatic" question. Since the diverse principles, methods, and subject matters, which beget the doctrinal oppositions among philosophies, are all plausible and equally primordial starting points for philosophical thought, *the* truth cannot be found. What can be found is a truth relative to a self-contained philosophy.

Radical or *de jure* philosophical pluralism presupposes that no set of substantive principles, or any method, or subject matter can rationally preempt its alternatives. Thus each philosophy, because radically diverse, is self-contained. And, therefore, no philosophy can be correctly interpreted or soundly criticized in terms of another philosophy's (alien) principles or methods.[12] Any meaningful criticism of a philosophy, which goes beyond the jejune detection of logical fallacies and contradictions, can only be internal to that philosophy: *its* method has not been consistently applied; *its* principles have not been clearly stated; *its* subject matter has not been adequately explored or enlarged; *its* tenets have not been systematically interrelated. But clarifying, developing, and systematizing a philosophy, are evidently the tasks of a disciple not of an adversary of that philosophy.

995a27-33), see Joseph Owens, *The Doctrine of Being in the Aristotelian Metaphysics*, 2d ed., rev. (Pontifical Institute of Mediaeval Studies: Toronto, 1963), pp. 211-219.

[12] See Walter Watson, *The Architectonics of Meaning: Foundations of the New Pluralism* (Albany: State University of New York Press, 1985), p. ix: ". . . the attempt to establish the one true philosophy by refuting all other philosophies is not destined to succeed, for the refutation of all other philosophies depends on interpreting them in the terms of one's own philosophy . . . ."

What task, though, is left to the philosophical pluralist whom we may suppose to be, by definition, neither disciple nor adversary of other men's philosophies. His or her task, as McKeon puts it, is to interpret and clarify the semantics of different modes of philosophical discourse by locating the common topics, hypotheses, and themes, and schematizing the different categories, facts, methods, and principles to which diverse philosophies appeal.[13]

Especially in a contemporary context, the notion of radical or *de jure* philosophical pluralism seems indistinguishable, or at least not very easily distinguished, from the notion of a conceptual relativism that regards the meaning and truth of sentences as radically dependent upon autonomous conceptual contexts, frameworks, schemes or (to use the all embracing and ultimate metaphor) "worlds" that are themselves diverse, irreducible, and, therefore, immune to external criticism.[14] If these conceptual worlds are ultimately self-contained, rationality itself, in any material sense, must be regarded as radically context-dependent.

This notion of conceptual relativism, which was first promoted by philosophers of science (Kuhn, Quine, and Feyerabend) about the history of scientific revolutions and subsequently generalized and

---

[13] See McKeon, "Discourse, Demonstration," p. 45.

[14] Cf. McKeon, "Dialogue and Controversy," p. 114: "Truth is perceived in perspective, and perspectives can be compared, but there is no overarching inclusive perspective"; Nelson Goodman, *Ways of Worldmaking* (Indianapolis: Hackett Publishing Company, 1978), p. 17: ". . . truth cannot be defined or tested by agreement with 'the world'; for not only do truths differ for different worlds but the nature of agreement between a version [of worldmaking] and a world apart from it is notoriously nebulous." For a succinct overview, see the Introduction to *Scientific Revolutions*, ed. Ian Hacking (Oxford: Oxford University Press, 1981), pp. 1-5.

popularized (notably by Richard Rorty) about the history of every mentalité,[15] turns aside, as so much labor lost, the classical pursuit of western philosophy.[16] Conceptual relativism rejects the pursuit of the common *logos* in favor of what Heraclitus called "private understanding,"[17] or what are today called "incommensurable conceptual schemes."[18] The schemes

---

[15] Cf. Richard Rorty, *Contingency, Irony, and Solidarity* (Cambridge: Cambridge University Press, 1989), p. 7: ". . . the human self is created by the use of vocabulary rather than being adequately or inadequately expressed in a vocabulary . . . truth is made rather than found."

[16] For a lively synopsis of the recent debate, diagnosis of the systemic Cartesian disease, and a suggested hermeneutical cure, see Richard J. Bernstein, *Beyond Objectivism and Relativism: Science, Hermeneutics, and Praxis* (Philadelphia: University of Pennsylvania Press, 1983).

[17] See Heraclitus, fr. 2.

[18] For a famous but densely argued attack on the conceptual scheme/world dualism, see Donald Davidson, "On the Very Idea of a Conceptual Scheme," *Proceedings of the American Philosophical Association*, 47 (1973-74), 5-20; reprinted in *Relativism: Cognitive and Moral*, ed. Jack W. Meiland and Michael Krausz (Notre Dame and London: University of Notre Dame Press, 1982), pp. 66-79. Diverse conceptual schemes are best understood as different languages that are not intertranslatable but which, nonetheless, are thought individually to either (1) organize or (2) fit the world. Among Davidson's salient objections are: (1) one can only "organize" that which is already individuated "according to familiar [linguistic] principles" and hence such schemes can be translated (p. 74); (2) the truth of a sentence is determined (*à la* Tarski) by the totality of sentences in the language in which it occurs ("Snow is white" is true if and only if snow is white") not by reference to something extra-linguistic. Truth, so conceived, cannot be intelligibly applied to an allegedly non-translatable language: see p. 76. For an argument, *contra* Davidson, that "interpretability" and *not* putative "non-translatability" is the criterion that allows one to establish the alternativeness of a conceptual scheme, see Nicholas Rescher, "Conceptual Schemes," *Midwest Studies in Philosophy*, Vol. 5: *Studies in Epistemology*,

are incommensurable because the one cannot be translated into the other. But radical philosophical pluralists portray the same kind of conceptual incommensurability; the history of philosophy is a series of conceptual revolutions or so-called paradigm shifts.[19] Yet, according to McKeon, philosophers necessarily and rightfully live in diversely ordered conceptual worlds: "Neither existence nor experience is *ab initio* an ordered whole of constituted facts or of spaced or sequential events."[20] What constitutes the "facts" as well as their conceptual order and meaning is the philosophers complex selection of a *method* with its attendant form of *interpretation*, of a *subject matter* or *selection*, and of *principles*.[21]

II. *"Evolutionary Thomism"*

As the sub-title of his book indicates, McCool treats "Thomism" as a nineteenth-century organism (which *Aeterni Patris* erroneously identified as the best specimen of a supposedly unitary scholastic philosophy)

---

ed., Peter A. French, Theodore E. Uehling, Jr., and Howard K. Wettstein (Minneapolis: University of Minnesota Press, 1980), pp. 323-345.

19 See McKeon, "Philosophy and Method," p. 669: ". . . the major steps in the history of philosophy seem to occur not by reason of adequate analysis and valid refutation of philosophies before their abandonment, but by simple decision to use some other method on some other subject-matter supported by critical propaedeutics, schematic comparisons, and historical reviews"; "The Methods of Rhetoric and Philosophy: Invention and Judgment," in *Rhetoric*, pp. 56-65; "Periodical revolutions occur in the history of philosophy which change the perspectives of philosophical inquiry and transform the meaning of philosophic words and the natures of things meant" (*ibid.*, p. 56).

20 McKeon, "Discourse, Demonstration," p. 50.

21 See Richard McKeon, "Philosophic Semantics and Philosophic Inquiry," in *Other Essays*, pp. 242-256.

that evolved into different twentieth-century species. The Thomist evolution occurred necessarily, McCool alleges, because of "the historical nature of thought itself" (p. 211).[22] But as the reward of evolutionary adaptability is survival, so too the evolutionary plurality of twentieth-century Thomisms allowed Catholic philosophy and theology "to maintain their hold on truth" (p. 211).

McCool's evolutionary metaphor, whether or not it is deliberately ambiguous, prompts an initial question about these twentieth-century "Thomisms." Darwinian evolution, supposing that is the root metaphor being used, requires that one biological species evolves into another species so that something specifically *new* emerges that cannot be classified as merely a version of the old. We should ask, then, whether everything that calls itself "Thomism," in whatever century, actually retains "St. Thomas's own fundamental principles" (p. 228) or do we, indeed, find that there has emerged a variety of new and distinctly non-Thomistic philosophical species? That McCool's metaphor should indeed be given a Darwinian interpretation, at least in regard to the *non-Thomistic* character of Maréchal's and Rahner's respective philosophies, I have argued elsewhere.[23]

McCool, nonetheless, speaks about "the legitimacy of systematic pluralism" (p. 229) but what he means would be expressed more accurately, I think, by

---

[22] McCool directly attributes this thesis to Bouillard but, once again, it seems to characterize, indirectly, his own point of view.

[23] See Denis J. M. Bradley, "Transcendental Critique and Realist Metaphysics," *The Thomist*, 39 (1975), 631-667, "Rahner's *Spirit in the World*: Aquinas or Hegel?" *The Thomist*, 41 (1977), 167-199; "Religious Faith and the Mediation of Being: The Hegelian Dilemma in Rahner's *Hearers of the Word*," *The Modern Schoolman*, 55 (1978), 127-146.

the phrase "legitimate philosophical pluralism."[24] The capacity "[to] preserve the fundamental meaning of the Christian mysteries" (p. 230) determines, apparently, whether a given philosophy is or is not "legitimate" so that "legitimate philosophical pluralism" is the class of all such Christian or revelation-congruent philosophies. Accordingly, McCool's second underlying thesis is that *legitimate* philosophical pluralism and not the multiple species of evolved Thomism guarantees our hold "on the invariant truth of the Christian faith" (p. 228).[25]

Given this second thesis, it would be useless here to dilate over who is or who is not genuinely "Thomistic" among twentieth-century practitioners of Thomism. The accolade, since it is philosophically evaluative as well as historically descriptive, is always liable to dispute. How "Platonic" were the neo-Platonists? Anyway, we can allow, in today's more latitudinarian ecclesiastical scene, that there are respectable schools of "Neo-Scholastics" without feeling any pressure to authenticate "Neo-Thomists" as faithful disciples of Aquinas.

Nonetheless, it is expedient to give a peremptory answer to the quite different question, whether a "systematic pluralism" of presumably legitimate but *non-Thomistic* philosophies is possible or even necessary, as McCool seems to imply (pp. 228-229), "on the basis of St. Thomas' own epistemology and metaphysics" (p. 2). It is this latter claim, which is a corollary of McCool's second thesis, that I find both surprising and ill-

---

[24] Cf. McCool, p. 229: "The [Second Vatican] Council's commitment to 'a philosophical tradition perennially valid' does not deny the legitimacy of a plurality of speculative systems through which that philosophy is mediated."

[25] Cf. McCool, p. 211: "It is one thing for Gilson to have established that historical development and pluralism existed as a *matter of fact* in the Christian philosophy of the Middle Ages, but Gilson never claimed that they are *necessary on principle* if philosophy and theology are to maintain their hold on truth."

conceived: not that different and non-Thomistic philosophies have been and, therefore, can be legitimately used to express revealed truth, for that is a conclusion that Gilson and other medievalists have shown to be inescapable given the precedent set in the Middle Ages, but that St. Thomas's *own* epistemology and metaphysics can be the legitimate basis for different philosophies.

McCool, however, reports Bruno de Solages' argument that at least one Thomistic *principle*, the "analogy of being," supports the notion that Aquinas's own epistemology and metaphysics license non-Thomistic philosophies. Bruno de Solages' assumption is that no set of concepts can adequately represent reality, and, hence, all philosophies using concepts inadequately represent reality. By adding yet another assumption, that the concepts of every philosophy are "both alike and unlike the reality that they represent" (p. 214), Bruno de Solage (and perhaps McCool?) concludes that each philosophy's inadequate conceptual grasp of reality is an *analogous* "representation of reality" (214).

Still, what does this Suarezian notion of mediate or imperfect conceptual *representation* of the sensible world have to do with the Thomistic doctrine about the analogous *meaning* of terms? Aquinas's doctrine of analogy (i.e., that part which is pertinent to Bruno de Solages' claim) is a theory about how predicates, whose original meanings are derived from creatures, may be applied to God by knowers having immediate cognition of sensible beings. In its Thomistic context, that "being" or any other metaphysical perfection is said analogously is a *conclusion* derived from considering the causal relationship between creatures and God. God, because he is the cause of creatures, must be said (not altogether equivocally) to possess the perfections of creatures, but since God is His being, these perfections must be predicated of God non-univocally, only by way of eminence and negation. Thus the meaning of these predicates, when applied to God, is neither exactly the same nor entirely different than when applied to creatures.[26]

Let us return to the main point, the provenance of philosophical pluralism as McKeon conceives it. If diverse philosophies are constituted by their principles, methods, and subject matters (and what else could they be constituted by?), then *only* St. Thomas's own epistemology and metaphysics can rest on *his* noetic and metaphysical principles. Of course, one might apply Thomistic principles and methods to a new subject matter, and then we might well acknowledge that a disciple has developed a new Thomistic philosophical science, say "Thomistic philosophical anthropology". But this, surely, is not the kind of philosophical proliferation that any pluralist will find significant or worthy of celebration. Radical philosophical pluralists champion the uncompromising epistemological and metaphysical diversity that engenders contradictory philosophical tenets and not the genial doctrinal nuances that color members of a school.

De jure philosophical pluralism does not sustain the idea of "Thomism" or any other philosophy organically "developing" into new and distinct philosophies. New and distinct philosophies are autonomous creations, self-produced, as it were, by the diverse philosophical methods and principles that contextualize the initially ambiguous meaning of common topics and themes.[27]

---

[26] Cf. *ST*, I, q. 12: Suárez, *III De Anima*, c. 2, n. 25 (Vivès, p. 1156): "Multum deficiens est talis conformitas sive unitas, sicut umbra vel pictura a realitate exemplaris deficit. Unde per quantum solummodo analogiam attributionis cognitio unitatem seu identitatem, ac proinde similitudinem cum objecto habet."

[27] Cf. McKeon, "Philosophy and Method," pp. 672-673: "The relations among philosophies are not simple differences concerning the same or comparable problems, nor can they be reduced to a translation formula which will transform a philosophic doctrine into the equivalent statement proper to another philosophy

McCool, however, seems prepared to embrace only a limited or (what I have labeled) a "legitimate" philosophical pluralism. This limitation doubtless arises from his extra-philosophical wish not to jeopardize "the fundamental meaning of the Christian mysteries when they must be expressed through historically conditioned concepts in a plurality of diverse systems" (p. 230). Yet, *de jure* philosophical pluralism cannot be limited by any appeal to external criteria. But the conceptual relativism that follows in the wake of unconditional *de jure* philosophical pluralism hardly allows the latter notion to be regarded as the *ancilla* of the Christian mysteries. Who, in such case, is either the master or the servant of *the* saving truth?

The challenge posed by *de jure* philosophical pluralism has not escaped the attention of Joseph Owens.[28] Owens, like McKeon, regards *de facto* philosophical diversity as inevitable. There are diverse starting points for philosophy which Owens identifies (again, like the early McKeon) as "three broad areas, things, thought itself, or language" (151). Although any of these is a "complexity" involving the other two,[29] the

---

. . . the patterns [among the structures of philosophic systems] are multidimensional and can be made to coincide only for limited areas."

[28]See Joseph Owens, C.Ss.R., *Cognition: An Epistemological Inquiry* (Houston, Texas: Center for Thomistic Studies, 1992); Introduction to *Towards A Christian Philosophy* (Washington, D.C.: The Catholic University of America Press, 1990), pp. 1-59; "Aquinas and Philosophical Pluralism," in *Thomistic Papers II*, ed. Leonard A. Kennedy, C.S.B. and Jack C. Marler (Houston: Center for Thomistic Studies, 1986), pp. 133-158; "The Primacy of the External in Thomistic Noetics," *Église et Théologie*, 5 (1974), 189-205; "Reality and Metaphysics," *Review of Metaphysics*, 25 (1971-1972), 638-658.

[29] Cf. Owens, "Reality and Metaphysics," p. 647: "The combined complexities of things, thought, and language are accordingly present in the immediate object from which metaphysics takes its start."

history of philosophy is the story of what happens when a philosopher takes one or the other of these three complexities as his or her starting point. The lesson that Owens draws from the story, as did Gilson before him, is that the starting point of a philosophy inevitably dictates its conclusions. But this lesson, to be sure, must be carefully distinguished from the conclusion that grounds *de jure* philosophical pluralism: that, rationally, all starting points are equally primordial or valid or useful for philosophy. On this issue, Owens resolutely parts company with McKeon.

Owens acknowledges that the history of philosophy leads only to an *historical relativism* about the ultimate priority of any starting point for philosophy. Once it has chosen its own principles, each different philosophy is self-contained. Although it can criticize another philosophy, but only dogmatically on the basis of its own principles, a coherent philosophy cannot, using its peculiar principles, ground a rational transition to another philosophical standpoint. Yet, a *philosopher* sometimes abandons his or her principles if the conclusions that follow from them become "unacceptable." But to abandon a self-contained philosophy is for a philosopher to make a "quantum jump into another philosophical orbit."[30]

At this point, one can imagine the radical philosophical pluralist, who would happily confirm the notion of philosophers making logically discontinuous "jumps" from conceptual orbit to conceptual orbit, encouraging Owens to admit that the radical diversity of philosophies entails *de jure* philosophical pluralism.[31] But Owens is quick to eschew a "thoroughgoing type of relativism," or *de jure* philosophical pluralism, that would allow that each of the possible beginnings for

---

[30] Owens, *Towards A Christian Philosophy*, p. 40.

[31] Cf. Rorty, p. 6: ". . . the world does not tell us what language games to play . . . ."

philosophy (i.e., things, thought, or speech) to be equally plausible and primordial.[32] In his numerous publications, Owens has reiterated and defended the *de jure* primacy of the Aristotelian starting point for philosophy: *sensible things* "existing in themselves outside our cognition."[33] Against any kind of attempt to move from ideas to things, Owens contends that "nothing can be more certain than the tenet of their [sensible things] existence."[34] Perception first grasps not modifications of the percipient but "distinct objects complete in themselves."[35]

Owens' strategy in defending the Aristotelian starting point, to put it in general terms, consists in arguing that the other possible starting points pose unresolvable problems of "how to get from one to the other"[36] complexity. Beginning with thought, we cannot validly construct an epistemological bridge to really existing things that are not mere appendages to thought.[37] Beginning with speech, which has no fixed signification, we cannot reach "stable thought and things."[38] But by beginning with "the first objects of which one is directly aware,"[39] sensible things, we can adequately show that our knowledge of thought and speech has "its ultimate epistemological source in the

---

[32] Cf. McKeon, "Discourse, Demonstration" p. 45: "There is no pre-established priority of being, cause, or rule among things, thoughts, actions, and statements; each in turn may be made fundamental in deliberation or judgment or demonstration."

[33] Owens, *Cognition*, p. 53.

[34] *Ibid.*, p. 236.

[35] Owens, "Primacy of the External," p. 195. Cf. *ibid.*, p. 201: ". . . sensible things in their real existence are always prior."

[36] *Ibid.*, p. 329.

[37] See Owens, "Primacy of the External," p. 197.

[38] *Ibid.*, p. 334.

[39] *Ibid.*, p. 321

sensible things themselves."[40] In short, the *primacy*, *virtuality*, *penetration*, and *range*, as well as the *certainty* of the Aristotelian starting point can be shown to be greater than any of the other starting points.[41]

Nonetheless, if *de jure* philosophical pluralism is to be rejected, at least by anyone who consistently adheres to the Aristotelian starting point, what explains, finally, *de facto* philosophical pluralism? Owens falls back on what is admittedly a commonsensical observation: people are physically and psychologically "built differently."[42] My global philosophical choice to begin with things, or thought, or language, and my particular philosophical choice to begin with any one of the myriad particular starting points contained in things, thought, or language, depends, Owens thinks, "in large part [my] temperament and upbringing."[43]

This commonsensical explanation, however, has an obvious defect: it explains the diversity of philosophies by appealing to non-rational causes and it does not adequately address the question whether philosophical *reason* is itself inherently diverse in the choice of its starting points. Owens' explicit explanation of philosophical diversity, abandons the terrain of philosophy for physiology or psychology, or more plausibly, for cultural or intellectual history.[44] Yet,

---

[40] *Ibid.*, p. 334.

[41] McKeon, by contrast, argues that philosophical discourse provides the only common, non-controversial beginning that adequately *comprehends* the other starting points: "The intelligible structure of discourse is at once a structure of symbols related and transposed, of actions performed or projected, and of things signified and ordered. The frameworks of things, thoughts, actions, and symbols are assimilated in the common framework of discourse" ("Discourse, Demonstration," p. 51).

[42] Owens, "Aquinas and Philosophical Pluralism," p. 150.

[43] Owens, *Cognition*, p. 333.

[44] Cf. Richard McKeon, "Philosophy and Method," *The Journal of Philosophy*, 48 (1951), 656: "The subject-matter in

Owens himself obliquely mentions other criteria for his philosophically *reasoned* choice of the Aristotelian starting point for philosophy. He judges that starting point to be *primary*, and to have more *potentiality*, and to be more *penetrating*, *comprehensive*, and *certain* than the others.

These rational criteria, and others like them, are what Nicholas Rescher refers to as "epistemic values" or "cognitive values" that ground the fundamental commitments of a philosophical "system."[45] Rescher's important book deserves careful study not only by Thomists or any other philosophers who seek to maintain the purported epistemological priority of their own starting points but by any theologian considering the role of philosophy as *ancilla* to faith. *The Strife of Systems* is a lucidly written, historically learned, tenaciously argued, and provocative defense of radical or *de jure* philosophical pluralism based on the irreducible plurality and diversity of cognitive values. Along with McKeon's, Rescher's discussion of unconditional *de jure* philosophical pluralism helps expose the problems inherent in McCool's theses about "evolutionary Thomism." But I shall not implicate McCool any further in this discussion of *de jure* philosophical pluralism which, to be fair, is ambient to the stated topic of his book. In the rest of this paper, I shall summarize the more important of Rescher's theses, suggest some of the replies to Rescher that the conscientious Thomist or any energetic would-be "realist" would need to develop in

---

terms of which the problems of philosophy are stated is determined roughly by the fashions of times and cultures."

[45] See Nicholas Rescher, *The Strife of Systems: An Essay on the Grounds and Implications of Philosophical Diversity* (Pittsburgh: University of Pittsburgh Press, 1985), esp. pp. 95-115. For an earlier and briefer statement, see Nicholas Rescher, "Philosophical Disagreement: An Essay Towards Orientational Pluralism in Metaphilosophy," *Review of Metaphysics*, 32 (1978), 217-251.

convincing detail, and, finally, broadly respond to some of the paradoxes engendered by Rescher's own position.

III. Cognitive Values

Rescher, unlike McKeon, makes no attempt to catalogue or schematically reduce the historically numerous subject matters that have been investigated, principles that have been developed, and methods that have been used by various philosophers; he is more interested in determining the criteria that enable a philosopher to make a *rational choice* of a method, a subject matter, and principles. Since philosophy deals with the "big questions" about "being and value in general" (p. 17), Rescher allows that any principle, or subject matter, and method, if logically consistent, is philosophically relevant. The rules of logic are, in any case, precise common norms for valid philosophical arguments. And in a rougher sense, there are common "facts of experience" which are drawn from common sense, science, everyday life, culture, received traditions, and history. But what is the philosophical *significance* of these facts?

"Realism," if we mean by that a philosophy that pretends to have an immediately coherent knowledge of the world in itself, is naïve: the world has multiple and, so the philosopher discovers, latently contradictory features that need to be resolved. Thus Rescher allows no indefeasible epistemological datum, but he does presuppose that philosophical reason is committed to comprehension and understanding, and, therefore, that it must render our beliefs *systematically* coherent. In attempting to do so, philosophers have methodically slaughtered every other philosopher's sacred cow. But the source of radical philosophical pluralism lies neither in undetected logical slips nor in factual inadvertency but in the peculiar cognitive values that guide each philosopher in rationally framing and defending his or her primitive set of philosophical tenets about the *significance* of the facts.[46]

Taken individually, common facts tentatively commit us to certain plausible cognitive claims which, taken together, prove to be inconsistent. Together these propositions constitute what Rescher aptly tags an "aporetic cluster" or set of theses which, when used individually as premises in valid arguments, lead to contradictory conclusions, i.e., to an explicit philosophical antimomy. Typically, philosophy is the effort to put our pre-philosophical cognitive commitments into a consistent and coherent *rational* order so that we can, subsequently, avoid antinomies by eliminating some and preserving other of the propositions in the original aporetic cluster. But the "plain facts," each of which supports a plausible thesis, do not determine how these theses are to be systematically ordered and, therefore, which are to be eliminated and which are to be preserved. The same (common) body of factual evidence can be evaluated differently by different philosophers. Should we stress the *uniform*, *established*, *general*, but *superficial* reliability of perception or those *idiosyncratic*, *illuminating*, scientifically *interesting*, *novel*, and *important* perceptual anomalies? My cognitive values, which are themselves open to reconsideration and

---

[46] Rescher's complex "sampler of cognitive values" (p. 102) has four quadrants: (1) *individual values* which are (a) utility analogous or *significance-oriented values* (important, significant, illuminating, informative, interesting, and weighty, and their opposite disvalues) and (b) probability-analogous or *plausibility-oriented values* (plausible, relevant, uniform, established, ordinary, general, natural, simple, and exact, and their opposite disvalues; and (2) *contextual values* which are (a) *pragmatic values* (central, immediate, firm, fundamental, fruitful, deep, urgent, and pressing, and their opposite disvalues) and (b) *systemic values* (coherence, uniformity, comprehensiveness, completeness, self-supportiveness, economy, elegance, and regularity).

change, direct my decision as to which facts, among the plethora of facts, are significant, important, and central.

De jure philosophical pluralism, which carries with it ineradicable debates and controversies, is rooted in the overdetermined or aporetic character of common experience and in the philosopher's choice of a rationally underdetermined but rationally preferred set of theses which he selects, guided by his cognitive values, from the original aporetic cluster. On the basis of the theses selected, the philosopher will avoid antinomies. Although there is always more than one logical exit from any antinomy, their number is finite, and so, typologically, there emerge schools of philosophy rather than a myriad of unique philosophies.

In Rescher's account of philosophy, everything turns on the latently contradictory character of our original experience of the world. In both senses of the verb, philosophy *invents*--it discovers and makes--contradiction.[47] Our central philosophical concepts are, first of all, borrowed and, subsequently, can never be unconditionally divorced from the ordinary concepts used to coordinate the facts of everyday and scientific experience. But everyday concepts are more pragmatically efficient than precise. And while scientific concepts are precise, they rest on empirical hypotheses that are provisional and subject to change. Accordingly, neither the underdetermined facts of everyday life nor the precisely but putatively determined facts of science can settle which of the elements fused in the ordinary concept are "ultimately determinative or decisive" (p. 47) for systematic thought. Philosophy, which is systematic thought, is the attempt to make that determination.[48]

---

47 Rescher, p. 63: "For it is bound to happen that, whenever we set down the aggregate of 'natural' theses that expound what makes intuitive good sense in any area of consideration, the results prove to be inconsistent."

48 Accepting or not accepting a set of theses involves "weighing the costs of abandonments against the benefits of

For Rescher, the ambiguities and imprecision of everyday usage, and the tacit and revisable assumptions undergirding the precisions of scientific language, stand in contrast and inevitable conflict with the even more precise "scientific aspirations" of philosophy: philosophy strives for "absolute universality and precision in the delineation of concepts" (p. 49). But such clarification comes at high price; it is at once cure and disease. In exposing their ambiguous meanings, the philosopher sunders the contingent but pragmatically effective unity of fact-coordinated concepts, *invents* the aporetic character of experience, and thus explicitly opens philosophy to the rational antinomies that so threaten the coherency of reason. But antinomy resolving necessarily forces the philosopher, whom Rescher romantically describes as the "tragic" champion of "abstract rationality" (p. 54), to espouse cognitive values that are not abstract and universal but are concrete and personal since they are the result of his or her experience.[49] Yet, it is precisely his or her *personal* set of cognitive values that allow the philosopher to winnow some of the propositions from the original aporematic cluster so as to avoid the antinomies that would ensue if all plausible theses were held at once.[50]

While cognitive values are rational, they are not themselves objective "facts"; they are the subjective evaluative "inputs" (p. 105) that we bring to the facts.

---

retentions" (p. 109) *vis-à-vis* the philosophical system that will eventuate from these theses. Joseph Wayne Smith, "Against Orientational Pluralism in Metaphilosophy," *Metaphilosophy*, 16 (1985), 214-228 accuses Rescher of circular reasoning: we can assess the systematic consequences of a thesis only if we first have pre-systematic criteria for sorting out competing theses. Rescher, however, does not claim that the systematic consequences of a set of theses are the sole criteria for accepting or rejecting themn: see p. 133.

[49] See Rescher, p. 131.

[50] See Rescher, p. 109.

Since not all cognitive values can be simultaneously and equally affirmed, my (or my philosophical school's) set of operative cognitive values is always subject to challenge.[51] My cognitive values reflect "my course of experience" (p. 136): my nature but more importantly my intellectual nurture, my personal interests, and my response to current cultural beliefs and issues.

Rescher holds that a philosopher, if he or she does not wish to speak solely to the saved, can defend the principles of a philosophical system only by "exoteric" arguments that are "system-external." Exoteric arguments always depend on examples and plausible *analogies* to "cognate situations that seem unproblematic" (p. 99) in the common or extra-systematic arena. But arguing from analogies involves rationally disputable judgments about the *degree* and *extent* of the analogy, i.e., the weight or importance of its relative similarities and dissimilarities with the philosophical issue under examination. Consider, for example, the endless possibility for disputing what should count as the prime instance of pre-philosophical "knowledge" that can then serve as the archetype for philosophical knowledge. Is it common sense, everyday empirical knowledge of the world, mathematics, the physical sciences, religion, law, art, or literature? These all too familiar candidates keep appearing and disappearing on the philosophical stage.

Although Rescher locates the source of *de jure* philosophical pluralism in diverse cognitive values, which antedate and ground the explicit choice of a philosophical method, set of principles, or subject matter, he and McKeon are in agreement on the fundamental issue: "No unproblematic, universally conceded basis of first principles is available."[52]

---

[51] See Rescher, p. 126.

[52] Cf. McKeon, "Philosophy and Method," p. 672: "The subject-matter of philosophy is universal, and there is no reason *a*

Owens, however, argues that there is an indefeasible, initially unproblematic, and certain basis in the immediate *first-order* judgment that "sensible beings really exist."[53] This basis is not, of course, universally acknowledged as the proper starting point for philosophy and, indeed, it may itself be assaulted by *second-order* philosophical doubts and queries.[54] But Owens contends that second-order problems are in principle resolvable and are, in any case, never sufficient to overthrow our first-order or pre-philosophical certitude about existing sensible beings.[55]

Still, Owens himself concedes much of what McKeon and Rescher claim about philosophical orientation: changes of philosophy are "ideological reorientation[s]" or "quantum jumps to a different value outlook" (p. 121). They can be rationally legitimated *ex post facto*, not in terms of the old but only in terms of the newly accepted cognitive values. Thus McKeon, Owens, and Rescher concur that external arguments against a philosophical "system" inevitably fall back on alien philosophical standards.[56]

---

priori why any starting point should provide better principles than any other . . . ."

[53] Cf. Aquinas, *De Ver.*, q. 1, a. 1: ". . . quod primo intellectus concipit quasi notissimum, et in quo omnes conceptiones resolvit, est ens . . . ."

[54] See Owens, *Cognition*, p. 235: "'Something really exists' . . . cannot be overthrown when attention is focused solely upon itself. As a first order judgment, aside from any philosophical framework, it turns out to be resistant to Skeptical shattering."

[55] See Owens, *Cognition*, pp. 237-238.

[56] See Rescher, p. 121; Owens, *Cognition*, pp. 333-334; McKeon, "Dialogue and Controversy," p. 122.

IV. *Vade Mecum* of a Would-Be Realist

Owens' Aristotelian or "realist" starting point[57] (the immediate and certain judgment that sensible beings exist)[58] represents what Rescher, borrowing a term from William James, calls "absolutism" (p. 142), a starting point that is allegedly objective, universal, and rationally compelling for--let us say--all "ideal philosophers."[59] Rescher's "orientational pluralism," however, holds that any "cognitive perspective" (p. 176), absolutism included, necessarily rests on selected cognitive values or "a particular, and highly problematic, probative orientation" (p. 180).[60] For example, if a philosopher

---

[57] For Owens' reluctant use of the term "realism", which he regards as tainted by a post-Cartesian bias in favor of the priority of cognitional existence over the immediate cognition of the existence of sensible beings, see *Cognition*, pp. 325-326. Hilary Putnam, *The Many Faces of Realism* (La Salle, Illinois: Open Court, 1987), p. 4 acknowledges that "realism", in its present-day scientific and common sense usage, "can be claimed by or given to at least two very different philosophical attitudes (and, in fact, to many)."

[58] See Owens, *Cognition*, p. 237: "Once the epistemological priority of the judgment of real [sensible] existence has been satisfactorily established, the judgment itself is shown to be unassailable by philosophical frameworks that depend upon it for their own existence and functioning"; p. 239: ". . . all the philosophical frameworks can be explained on the basis of derivation from the perception of really existent sensible things."

[59] Rescher mentions Kant, Ayer, and Reichenbach as exemplary "absolutists": see p. 179. Cf. Owens, "Aquinas and Philosophical Pluralism," p. 144: "With Aquinas the norm is the sensible thing existent in itself . . . Quite as in Aristotle, there is found in the sensible world the firm basis for common and absolute truth."

[60] Owens stresses "priority" and "immediacy": "The origin of human cognition was found to lie in really existent sensible things. These are the *first* sensed, the *first* perceived, the *first* known" (*Cognition*, p. 236; italics mine). But are the

stresses the significance of the unusual rather than the ordinary, the allegedly universal objectivity and immediacy and coherence of our everyday knowledge of sensible beings can always be shown to be defeasible in certain instances.[61] The Aristotelian realist, by contrast, must argue that such anomalies are odd and parasitical; one can only make sense of them by reference to the stabilities of our ordinary knowledge.

Rescher, however, is more interested in the relationship of philosophy to science than to common sense.[62] "Realism" becomes globally doubtful once we move beyond the pragmatically secure but grossly imprecise and superficially explanatory claims of common sense to the more precise, deeply explanatory, but certainly defeasible claims of cutting-edge science. Contemporary science has realist aspirations as is evidenced in the relatively stable certitudes of its schoolbook redactions; but, in state-of-the-art theorizing, science can only asymptotically and approximately approach the truth about how the cosmos should be described and explained. Put simply, today's scientific theory, although it may be the best that we can do today, contains in all likelihood tomorrow's errors, but, in turn,

---

absoluteness of these Rescherian "values" called into question by Owens' own remark that "the choice [of a starting point in philosophy] will in practice depend in large part on temperament and upbringing" (*Cognition*, p. 333)?

[61] Cf. McKeon, "Discourse, Demonstration," p. 50: "Occurrences, observations, errors, illusions, and fictions are all facts; and errors are sometimes as illuminative of experience as controlled, well-grounded experimental observations."

[62] Cf. Rescher, p. 49: "The philosopher has scientific aspirations . . . His aim is to articulate what is so always and everywhere, flatly and unqualifiedly, absolutely and unconditionally."

tomorrow's truths will be corrected the day after tomorrow.[63]

Fortunately, we can leave realist philosophies of science to fend for themselves.[64] On either realist or anti-realist views of science,[65] one must ask what is the status of orientational pluralism itself? Rescher states (p. 178) that orientational pluralism is a descriptive

---

[63] See Rescher, pp. 211-212. For the details of Rescher's view of science, see his magisterial *A System of Pragmatic Idealism*, vol. 1: *Human Knowledge in Idealistic Perspective* (Princeton: Princeton University Press, 1992).

[64] See Rom Harré, *The Principles of Scientific Thinking* (Chicago: The University of Chicago Press, 1970). For an account of the cumulative character of scientific knowledge, based on the "moderate realism . . . usually associated with Aristotelian Thomism," see William A. Wallace, "Causality, Analogy, and Scientific Growth," in *From a Realist Point of View: Essays on the Philosophy of Science* (Washington, D.C.: University Press of America, 1979), pp. 201-215; "The Intelligibility of Nature: A Neo-Aristotelian Approach," *Review of Metaphysics*, 40, no. 3 (1987), 535-557.

Karl R. Popper, "The Aim of Science," pp. 191-205, in *Objective Knowledge: An Evolutionary Approach* (Oxford: Clarendon Press, 1972), while disavowing "the essentialist doctrine of ultimate explanation" (p. 194), nonetheless, contends that scientific theories, especially since they can be falsified, are "genuine assertions about the world; for they can *clash* with something we never made" (p. 197). For a robustly realist use of the nominal/real essence distinction, see Harré, p. 196: ". . . the position of the realist, [is] that things and substances differ by their internal constitutions and that these are the real differences, discrete or continuous real differences, upon which the differences of manifested characteristics of things depend."

[65] How this issue is resolved may not be relevant to a realist critique of *de jure* philosophical pluralism: see J. Owens, C.Ss.R., "Our Knowledge of Nature," *The American Catholic Philosophical Association*, 29 (1955), 63-86, for an argument on behalf of the conceptual independence of Aristotelian natural philosophy, modern science (and thus, presumably, the philosophy of modern science), and Thomistic metaphysics.

metaphilosophical thesis about the methodology of philosophy, and hence is entirely neutral about how ontological or "furniture of the world" questions should be settled. Yet, he admits that orientational pluralism is also a "substantive philosophical position" (p. 182), or more precisely "a position in normative metaphilosophy" (p. 264), which is dependent upon and justifiable only in terms of its own prior set of cognitive values.[66] Rescher's substantive position, however, is "quasi-Kantian"[67] since cognitive values are subjective in origin; they are "inputs."

Rescher, to be sure, does not think that we simply "choose" our cognitive values; rather, we find them--like "our nature, personality, disposition, [and] predilections"--"ensconced in place" (p. 126). If so, what finally accounts for my *operative* set of cognitive values? Rescher falls back on a combination of non-rational and rational causes to account for why, at the deepest level, we think the way we do: "A person's cognitive values, like his or her values in general, reflect a diversified spectrum of causal determinants" or "conditioning experiences of every kind" (p. 130), with nurture playing a more decisive role than nature. Value-formation is not a mechanistic or one-way process: experience determines our cognitive values, but our cognitive values determine what in experience counts as significant.

A realist, however, should take another but (according to today's philosophical fashions) widely rejected approach. Even if experience and cognitive evaluation form a loop, Rescher seems not to have advanced us much beyond Owens's more homely

---

[66] Orientational pluralism is defensible on a "cost-benefit appraisal" (p. 265) that shows it to be superior to skepticism, which dissolves philosophy, and absolutism which, in any of its forms, the history of philosophy shows to be untenable.

[67] Cf. Rescher, p. 215: ". . . cognitive values are matters of allegiance rather than observation."

explanation that, subjectively, we are just "built differently." On the principles of orientational pluralism, Rescher must allow that a realist could and should argue for the *objective* origin of our cognitive values: that, ideally, they cannot be based primarily on the *experience* of the individual but that they are properly dictated by the subject matter being examined.[68] Rescher readily grants that the physical sciences appeal to "interpersonally universal data" (p. 213) but he insists that no such appeal is possible in philosophy. But that conclusion, as Rescher well knows, is dictated by his own stance, *viz.*, his initial view of the aporetic character of experience. By definition, however, a realist is someone who does not conclude that the "facts of the matter," important matters at that, are always "incapable of enforcing a unique solution" (p. 108).[69] The trouble, for the realist, lies with Rescher's notion of "clarifying the facts."

To be sure, the philosopher, in attempting to get at the 'facts', must purge or redefine the equivocations and *prima facie* contradictions of pre-philosophical discourse. But while philosophy is undeniably committed to clarifying matters, Rescher too tightly embraces the Cartesian notion that philosophy is primarily committed to the absolute or unconditional *clarification of concepts*, while admitting that this effort to achieve perfect theoretical clarity strains and eventually over-taxes the cohesive power of the original facts. Against this rationalist notion of conceptual clarification,

---

[68] Cf. Etienne Gilson, "Vade Mecum of a Young Realist," in *Philosophy of Knowledge*, ed. Roland Houde and Joseph P. Mullally (Chicago: J. B. Lippincott Company, 1968), p. 393: "The realist should always maintain . . . that to every order of reality there must correspond a certain manner of approaching and explaining this reality."

[69] The realist preserves, in Aristotle's terms, the distinction between a scientific demonstration that starts from true and primitive premises, and a dialectical deduction that start from "reputable opinions": see *Topics*, I, 1.

which *always* invents philosophical antinomies,[70] an Aristotelian would certainly object that the desire for perfect theoretical precision must be tailored to the subject matter being investigated. Absolute precision or univocity of meaning is required in mathematics, but not in the other sciences which admit of focal and secondary meanings (metaphysics) and meanings that hold for the most part (ethics). Imprecision, with no failure implied, is built into philosophical no less than pre-philosophical discourse. As in strictly practical discourse, philosophical language need not and cannot be any more precise than the subject matter being discussed requires or warrants.

V. *Animadversions*

In defending the subjectivity of cognitive values, Rescher leans very heavily on the *de facto* philosophical pluralism that emerges from the history of philosophy. But Rescher slides too easily from the 'fact' that one can *reconstruct* the history of philosophy as a series of antinomies[71] to the assertion that underlying these philosophical antinomies there is always an aporetic cluster of pre-philosophical theses which in turn reflect our original but contradictory *experience*. Here his bias about the contradictory character of experience is "demi-Hegelian."[72]

---

[70] See Rescher, p.137: "Clarificatory pressure upon fact-coordinated concepts *always* engenders antinomies" (italics mine).

[71] Cf. McKeon, "Has History a Direction?", in *Other Essays*, p. 142: "Since philosophic semantics applies to the idea of philosophy as well as the facts of history, a philosopher who goes to history for factual confirmation of his ideas usually has little difficulty in finding that the facts support him, since he takes to his history the same philosophic semantics as guided him in his philosophy."

[72] Cf. Aristotle, *Topics*, I, 1, 100b26-38 (rev. Oxford trans.): "For not every opinion that seems to be reputable actually

I say "demi-Hegelian" because Rescher's orientational pluralism, although it ecumenically concedes the rationality of the idealist standpoint, denies that factual and conceptual contradictions can be overcome by a higher and comprehensive rational *synthesis*. Antinomies are avoided by *eliminating* theses from the aporetic cluster. But Rescher sweeps away the realist with the same broom. Our subjective cognitive values, and not the absolute or objective determinacy of being, set the direction of one's philosophizing.[73]

Yet, orientational pluralism licenses the rational validity of *all* of the subjective, value-determined exits (among them the realist and idealist exits) from pre-philosophical, aporetic discourse. At this point, the descriptive and normative aspects of *de jure* philosophical pluralism begin to clash. Rescher, however, denies "categorically and with vehemence" (p. 149) that orientational pluralism logically (or psychologically) entails that an individual philosopher be indifferent to the truth of his or her peculiar theses. Indeed, the normative orientational pluralist firmly judges that both realism and idealism are wrong in claiming to reach universal philosophical truths.

On the basis of his own set of cognitive values, every Rescherian philosopher can legitimately assert that his philosophical position is "optimally acceptable" (p. 178) or even "uniquely correct" (p. 179). Nonetheless, externally viewed (from the descriptive metaphilosophical standpoint occupied by the orientational pluralist), all philosophical theses enjoy a "parity or status" including those that would

---

is reputable. For none of the opinions which we call reputable show their character entirely on the surface . . . ."

[73] Cf. Rescher, p. 235: ". . . evidential monism . . . contends that considerations of objective evidential rationality will decide in favor of one position. The whole history of philosophy speaks against this absolutistic view, as does a theoretical analysis of the nature of philosophical problem solving."

"categorically and with vehemence" deny the standpoint of orientational pluralism. But Rescher explains that it is philosophy as a community enterprise, not individual philosophers, which indifferently holds contrary theses. He admits that descriptive orientational pluralism is indeed "a form of relativism", (p. 196) but it is an epistemological or axiological relativism not an ontological relativism. What is relative (to his cognitive values) are a philosopher's views or ideas of the truth, his truth-claims, or his justifications of his subjectively "warranted assertions."[74] In contrast to such relativized truth claims, there remains the notion of ontological truth, or *"the real truth of the matter"* (p. 187).

Yet, Rescher's distinction between ontological and epistemological relativism is doctrinally vacuous since we have no access to "God's own truth" (p. 180), or to come down a peg, to "the Recording Angel" (p. 187). Nothing though is lost by God's or the angel's silence. Once he occupies his chosen standpoint, the philosopher hardly needs his doctrines to be confirmed by the Recording Angel: "The philosopher sees his own position as right and proper; everyone else's position is seen as just wrong" (p. 160). This sense of being "right" applies no less to the orientational pluralist than to any other philosopher.[75]

Rescher moves fearlessly between the external and relativist *description* of philosophical pluralism and the internal, personal, and monist *prescription* of philosophical pluralism.[76] He continually reassures us that nothing impedes easy passage from one standpoint to the other. But I am dazzled by these moves. How can Rescher, who spends a whole book showing the

---

[74] See Rescher, p. 195.

[75] Cf. Rescher, p. 235: "Orientational monism thus finds its own support as a philosophical position in exactly the sort of methodology it itself endorses; the identification of one particular alternative as optimal in the light of cognitive values."

[76] See Rescher, p. 264.

relativistic and thus invincible rationality of other cognitively evaluative standpoints, so easily suspend what he has learned when he assumes his own dogmatic standpoint, the monistic *normativeness* of orientational pluralism as providing the optimal method of philosophizing and the optimal reading of the history of philosophy,[77] a conclusion which allows him to exclude everyone else's monistic assertions as "just wrong"?[78] It is not just that "one recognizes that other [philosophical] positions continue to be seen as tenable on other evaluatives bases" (p. 235); one also knows that each of them "is optimal from a particular evaluative orientation" (p. 232). Rescher, in opposition to Hegel's

---

[77] The norm for the orientational pluralist is that we *necessarily* and, therefore, ought explicitly to "appraise [philosophies and doctrines] by assuming a determinate value posture" (p. 272). Despite what its progenitor claims, this norm seems to be *the* implication of of Rescherian philosophical pluralism: cf. Rescher, p. 276.

[78] McKeon allows that *philosophical inquiry* commits a philosopher to a peculiar method, set of principles, and perhaps subject matter. But *philosophical semantics* preserves the "diversified treatment of the common problems and subject-matters of philosophical inquiry rather than setting the different arts and philosophies in controversial opposition in which semantic differences mark off the rival possibilities of unique statements of truth and determination of values" ("Philosophy of Communication," p. 119). In short, once one has attained the standpoint of *de jure* philosophical pluralism, one cannot descend to the monistic asertion of any one standpoint as rationally optimal: "Discrimination, not assimilation or reduction, is the method of philosophy, and dialogue proceeds by exploring the varieties of arguments and doctrines and testing assertions by their contradiction, not by adjusting all doctrines to a preferred position and refuting those which will not fit" ("Dialogue and Controversy in Philosophy," p. 109). Rescher sometimes allows the orientational pluralist to speak almost as irenically as McKeon (cf. p. 125), but not quite: the "labors [of differently oriented philosophers] may be misguided, but they are not futile" (p. 136).

alleged "conflation" (p. 272) of the individual and communal standpoints, attempts to apportion his insights: he attributes the relativistic insight of orientational pluralism to "the standpoint of the community" and the optimal or monistic insight to the individual philosopher.[79] But one need not invoke the subtleties of an all-comprehensive Hegelian dialectic to recognize that this communal standpoint is but Rescher's conception of the community of philosophers, and hence his relativism is no less personal and optimal than his monism.[80] Doubtless, this animadversion vents "my" cognitive values. Nonetheless, here paradox halts me in my efforts to comprehend the split philosophical consciousness of the orientational pluralist.

---

[79] See Rescher, p. 159: "The difference between 'my philosophy' and 'the enterprise of philosophy at large' is crucial."

[80] Rescher, who repudiates "the God's-eye point of view" (p. 15), stands, like Luther (see p. 150), where he *chooses* to stand, so that he may personally, as a committed partisan of certain cognitive values, do "good philosophical work . . . at the substantive level" (pp. 266-267).

TRANSCENDENTAL THOMISM AND *DE VERITATE* I, 9

John F. X. Knasas

In his recent book, *From Unity to Pluralism: the Internal Evolution of Thomism* (1992), Fr. Gerald McCool, S.J., rings down the curtain on the neo-Thomist revival as it was energized by *Aeterni Patris* (1879). He bases his proclamation of the revival's demise upon the perception of an ironic truth, viz., the encyclical's injunction to return to Aquinas uncovered an Aquinas at odds with the Thomism of the drafters of the encyclical. That latter-day Aquinas is the Aquinas of Maréchal, Rahner, and Lonergan.[1] In my opinion, it would be tragic for present discussion in Catholic philosophical and theological communities to allow McCool's ironic truth to stand as the last word in the story of neo-Thomism. This century should close on a more accurate note than that. Whatever misconceptions the drafters of *Aeterni Patris* entertained about Aquinas, the opposition to a post-Cartesian subjective starting point in Thomistic epistemology and metaphysics is not among them. The drafters were on target in their understanding of Aquinas as an inveterate *aposteriorist*.[2]

---

[1] Gerald A. McCool, *From Unity to Pluralism: the Internal Evolution of Thomism* (New York: Fordham University Press, 1992), pp. 229-30. In an earlier work, McCool writes in a similar vein: "For [the sanctioning of a post-Cartesian subjective starting point in Thomistic epistemology and metaphysics] would become the outcome of the historical rediscovery and systematic development of the genuine possibilities of the Angelic Doctor's own thought." *Catholic Theology in the Nineteenth Century: the Quest for a Unitary Method* (New York: The Seabury Press, 1977), p. 235.

[2] By calling Aquinas an *aposteriorist*, I am not denying the elaborate structure of knowing powers in the human soul, e.g., external senses, common sense, imagination, agent and possible

Recently in an article for *The Thomist*, I evaluated the revisionist claim that Aquinas' understanding of the human intellect includes a crucial *a priori* dimension functioning as a constitutive factor in our consciousness of objects.[3] Among the texts considered were the following: *De Ver.* I, 4, ad 5m with its claim that the truth of the first principles by which we judge everything proceeds from the truth of the divine intellect as from its exemplary cause; *De Ver.* XXII, 2 with its claim of our implicit knowledge of God; *In IV Meta.* lect. 6 on the inborn nature of the first principle. Not considered, however, is *De Veritate* I, 9c. In this text Aquinas says, "the intellect knows truth [insofar] as it reflects upon itself." Maréchal, Rahner, and Lonergan

---

intellects. I simply mean that in relation to actual cognition, these powers are pure *conditions* for knowledge. The structure of the knowing power performs no constitutive role *vis-a-vis* the known object. The situation is somewhat like a lock whose structure, though admitting some keys but not others, does not modify or change the keys it does admit. In contrast, the *apriorist* views the knowing power as more than a condition for knowledge; it is also a *case* of knowledge. Here the structure of the knowing power does perform a constitutive role. For the *apriorist* the knowing power is more like a pencil sharpener whose structure may modify or change the pencils that it admits. In *Cahier V* of *Le Point de départ de la métaphysique*, Maréchal attempts to argue the *apriorist* approach on the basis that the knowing powers are natures and nature is a principle of finality. Vd., Joseph Donceel, *A Maréchal Reader* (New York: Herder and Herder, 1970), pp. 149-50. This finality is glossed as the dynamism of the knowing power to its formal object. The intellect employs its dynamism to infinite being to ". . . implicitly project the particular data in the perspective of this ultimate End, and by so doing *objectivates* them before the subject." *Ibid.*, p. 152. In my opinion, however, the nature-as-finality idea as applied to the intellect need mean only the intellect's ordination to *abstract* intelligible content from the real. By itself the idea fails to mean any projection of content upon the data.

3 John F. X. Knasas, "Transcendental Thomism and the Thomistic Texts," *The Thomist*, 54 (1990), pp. 81-95.

seize upon these words as Aquinas' undeniable nod to what has been called transcendental method.[4]

---

[4] Maréchal's comment on *De Ver.*, I, 9, is: "Hence, in order to become reflectively aware of the already implicitly possessed truth of its objective content, the intelligence must find, in its direct act, a 'proportion to the thing,' a something that refers it to reality in itself. But this can be known only if we know the nature and the internal finality of this particular act, and the nature of this act can only become manifest to us through the nature of the active principle from which it derives, that is, from the nature or the internal finality of the intellect itself, 'made to be conformed to things.' Therefore, according to St. Thomas, the objective value of knowledge would be formally revealed to the subject through the analysis *of its own* apriori *exigencies*, acting upon a present datum, that is ultimately, through the objective exigencies of the necessary *affirmation*." Trans. by Joseph Donceel, *op. cit.*, p. 208. In his "Aquinas: The Nature of Truth," *Continuum*, 2 (1964), pp. 69-70, Karl Rahner says: "The little importance that St. Thomas' *a priori* attributes to the attempt at a comparison between the object in itself and the apprehended object is indeed manifest from the fact that he writes in the *De Veritate* that truth exists in the intellect by means of the judgment about the thing in itself. However this truth, this proportion of the judicative act to the thing (*proportio ad rem*), is not known, in the ultimate analysis, by a reflection, by a look at the thing itself, as perhaps we would hope, but by a reflection of the intellect upon itself. We do not intend to give here a precise interpretation of this text of St. Thomas, on which an immense amount has already been written. Let us say only that the transcendental reflection upon the conditions for the possibility of knowledge, i.e., upon the *natura intellectus*, as St. Thomas expresses it, . . ., is precisely, according to our way of looking at it, what St. Thomas, in the place cited, calls the *supra seipsum reflecti* of the intellect." Finally, in his *Verbum: Word and Idea in Aquinas* (Notre Dame: University of Notre Dame Press, 1970), p. 75, Bernard Lonergan expresses his opinion about *De Ver.* I, 9: "I cannot take this passage as solely an affirmation of the reflective character found in every judgment. Not in every judgment do we reflect to the point of knowing our own essence and from that conclude our capacity to know truth. Rather, in this passage Aquinas subscribed, not obscurely, to the program of critical

Not only because of the notoriety of McCool's book, but also because of the philosophical stakes involved, I would like to assess the transcendentalist claim for *De Ver.* I, 9. In my opinion, the use of transcendental method is at odds with a realist philosophy. For the employment of the method culminates in a constitutive understanding of the human knower that forever renders doubtful any claims to know the real. Just as a pencil sharpener modifies the pencils that it admits, so too the human knower constitutes the object of consciousness. [5] In such a situation, the

---

thought: to know truth we have to know ourselves and the nature of our knowledge, and the method to be employed is reflection."

[5] Maréchal, Rahner, and Lonergan all regard the dynamism of the intellect towards Being as a constitutive factor for our consciousness of beings. Maréchal remarks: "As soon as the intellect, meeting an external datum, passes to the second act under the formal motion of this datum and the permanent impulsion of the natural appetite, we have a particular, positive determination subsumed under the universal form of being, which previously was only the framework of and the call for all possible determinations. An 'object' profiles itself before consciousness." Donceel, *op. cit.*, p. 170. Also, on judgment: "Considered as a moment in the intellect's ascent towards the final possession of the absolute 'truth,' which is the spirit's 'good,' [affirmation] implicitly (*exercite*) projects the particular data in the perspective of this ultimate End, and by so doing *objectivates* them before the subject." *Ibid.*, p. 152. ". . . the 'datum' represented in us was constituted *as* an 'object' in our mind through the judgment of *affirmation*." *Ibid.*, p. 161. "For the subject is really knowing as such only to the extent that he formally takes part in the edification of the object." *Ibid.*, p. 118. The same constitutive approach can be noted in Rahner. Speaking of the agent intellect, Rahner says: "Insofar as [the agent intellect] apprehends this material of sensibility within its anticipatory dynamism to *esse*, it 'illumines' this material, . . ." *Spirit in the World*, trans. by William Dych (New York: Herder & Herder, 1968), p. 225, also, p. 221. Likewise, "Because it is apprehended in this dynamic tendency of the intellect . . . the particular sensible thing is known as finite, i.e., as incapable in its limitation of filling up the space of this

knower is in real doubt as to whether the constitution produces a distortion or manifests things for what they really are. Any resolution of the doubt would depend upon some determination of the objectivity of the *a priori* rule of constitution. In Transcendental Thomism this determination takes the form of retorsion, or performative self-contradiction, viz., any doubt of the *a priori* rule employs the rule in its very exercise and so destroys itself.[6] But to my mind, the skeptic is on good

---

dynamism. Because of this comparing of the particular thing to the absolute and ideal term of knowledge, the particular thing appears as existent (concrete being) in relation to being." Rahner, "Aquinas: The Nature of Truth," p. 67. Finally, in *Insight: A Study of Human Understanding* (New York: Longmans, 1965), Lonergan appears to share the same constitutive approach found in Maréchal and Rahner. In fact at p. xxii, Lonergan expresses his intention to incorporate what Maréchal calls the finality of the intellect. Noteworthy are points found in Lonergan's discussion of the notion of being. First, abstraction is described as a provisional disregarding of the intellect's unrestricted objective of being (pp. 355-6). This suggests that being is an expanse against which things are initially profiled and from which we temporarily depart as abstraction focuses upon some feature. Likewise, judgment is understood as "an element in the determination of the universal intention of being." (p. 358) This seems to mean that each judgment is profiled against the notion of being. Such a move enables us to see the judgment as an "increment in a whole named knowledge." The move also sets the stage for wondering to arise once more and to lead to further judgments. In sum, McCool, in my opinion, correctly describes both Rahner and Lonergan as "Maréchalian epistemologists" and "Maréchalian metaphysicians," "Twentieth-Century Scholasticism," *The Journal of Religion*, 58 (1978), pp. 218-19. Finally, on how an *aposteriorist* would acknowledge the dynamism of the intellect, see John F.X. Knasas, *The Preface to Thomistic Metaphysics: A Contribution to the Neo-Thomist Debate on the Start of Metaphysics* (New York: Peter Lang, 1990), pp. 49-50.

[6] For Maréchal's key exercise of retorsion, see Donceel, *op. cit.*, pp. 215-17, 227-8; for Rahner, "Aquinas: The Nature of Truth," p. 69; for Lonergan, *Insight*, p. 352 on being as

grounds to insist that this phenomenon is just what one would expect if the rule is merely *a priori* and not objective at all. The skeptic's employment of a notion of the objective arises naturally from previous acquaintance with limited rules of constitution. Why may not the "universal" constitutive rule be actually limited too? In sum, retorsion is indecisive for purposes of deciding between distortion or objectivity.

So, if Aquinas is a practitioner of transcendental method, then logically he should relinquish all realist claims. This is a great price to pay. I intend to show that one ought to read *De Ver.* I, 9, in the line of a straightforward *a posteriori* position.

II

The topic of the article is whether truth exists in the sense power. Aquinas answers qualifiedly. Truth exists both in the intellect and in sense but not in the same way. The difference is that the intellect knows the truth it possesses but sense does not. Aquinas describes the intellect's knowledge of truth in the second paragraph of the *responsio*:

> [Truth] is in the intellect (1) as following the act of the intellect and (2) as known through the intellect. [Truth] follows the operation of the intellect insofar as the judgment (*iudicium*) of the intellect is about the thing according as it is. [Truth] is known by the intellect insofar as the intellect reflects upon its act, not only insofar as it knows its act, but insofar as the intellect knows the proportion [of its act] to the thing: which is not able to be known unless the nature of its act has been known (*nisi cognita natura ipsius actus*); which is not able to be known unless the nature of the active principle is known (*nisi cognoscatur natura principii activii*), which is the

---

unrestricted. On whether Aquinas uses retorsion to "indirectly prove" the first principles, see Knasas, *Preface*, pp. 51-2.

> intellect, in whose nature it is to be conformed to things; hence according to this the intellect knows truth, that it reflects upon itself (*supra seipsum reflectitur*).

After remarking that truth both exists in the intellect and is known by the intellect, Aquinas explains both points. On the existence of truth in the intellect, he says: "truth follows the operation of the intellect inasmuch as it belongs to the intellect to judge about a thing as it is." Aquinas discussed this doctrine back in article 3. Treating the question of whether truth is only in the composing and dividing intellect, he remarks:

> . . . the nature of truth is first found in the intellect when the intellect begins to possess something proper to itself (*aliquid proprium*), not possessed by the thing outside the soul (*res extra animam*), yet corresponding to it, so that between the two--intellect and thing--an adequation may be found.

Such a moment is the intellect's judgment: "when the intellect begins to judge about the thing it has apprehended, then its judgment (*iudicium*) is something proper to itself--not something found outside in the thing." But the intellect judges ". . . at the moment when it says that something is or is not." In article 3 (and consequently article 9), Aquinas uses the words "*iudicare*" and "*iudicium*" to designate the intellect's formation of propositions. *Contra Gentiles* I, 58, *Amplius. Propositionis* confirms this equation:

> Furthermore, in the case of a proposition formed by a composing and dividing intellect, the composition itself exists in the intellect, not in the thing that is outside the soul (*in ipso intellectu existit, non in re quae est extra animam*).

In speaking of the proposition as what exists only in the soul, this text also identifies what *De Ver.* I, 3c, called "proper to the intellect and not possessed by the thing outside the soul."

Yet this correlation of judging with the intellect's composing of a proposition fails to reveal the fundamental nature of the *secunda operatio intellectus.* Earlier at *In I Sent.* d. 19, q. 5, a. 1, ad 7m, Aquinas characterizes the enunciation, or proposition, as a sign (*signum*) of the *secunda operatio.* In contrast, the *secunda operatio* itself is what *respicit esse rei.* [7] What, then, is the fundamental nature of the second operation itself?

Suitable help comes from Aquinas' discussions of how God knows the truth of enunciables, especially as the enunciables bear upon singulars.[8] These discussions involve explaining God's knowledge of particular things. One can conclude that a grasp of the basic nature of the human intellect's second act lies in an understanding of its knowledge of the singular.

In abundant texts Aquinas tells how the human intellect knows the singular. The reiterated Thomistic opinion is that the human intellect knows the universal directly but knows the singular by a reflection to sense or imagination. [9] This reflection should, then, be what Aquinas understands as the "composing" of the intellect's second operation itself.

---

[7] ". . . prima operatio respicit quidditatem rei; secunda respicit esse ipsius. Et quia ratio veritatis fundatur in esse, et non in quidditate, ut dictum est; ideo veritas et falsitas proprie invenitur in secunda operatione, et in signo ejus quod est enuntiatio, . . ." *In I Sent.* d. 19, q. 5, a. 1, ad 7m.

[8] *In I Sent.* d. 38, q. 1, a. 3 and *De Ver.* II, 7. The title of the *De Veritate* article is: "Utrum Deus cognoscat singulare nunc esse vel non esse, . . . ; et hoc est quaerere, utrum cognoscat enuntiabilia, et praecipue circa singularia."

[9] *In II Sent.* d. 3, q. 3, a. 3, ad lm. Also *De Ver.* II, 6c.

The *Contra Gentiles* once again provides a double-check. At *Contra Gentiles* II, 96, *Palam*, while showing that angels do not draw their knowledge from sensibles, Aquinas mentions as a contrast the human intellect's twofold operation. Of the second operation he says: "[Noster intellectus] componit autem aut dividit applicando intelligibilia prius abstracta ad res." The thought is quite clear: our intellect composes or divides by applying previously abstracted intelligibles to the thing. The composing of the intellect's second operation is its reflective reintegration of its knowledge of a commonality with its knowledge of an instance from which it drew the commonality.[10]

These reflections allow one to distinguish in Aquinas' thought a cognitional activity sense of judgment, i.e., what I have just elaborated as the basic nature of the *secunda operatio*, from a propositional sense of judgment.[11] The intellect's "composing" is

---

[10] Back at *De Ver.* II, 6, ad 3m, man's reflective knowledge of the singular is presented in terms strikingly similar to *C.G.* II, 96: ".. . . ideo potest applicare universalem cognitionem quae est in intellectu, ad particulare."

[11] ". . . 'judgment' has two meanings that require careful distinction. In one meaning it is the dynamic intellective act by which synthesizing existence is being grasped. In the other meaning, it is the static, frozen representation of that action's cognitional form. In the first meaning, it denotes the 'second operation' of the intellect. . . . In the first meaning, the object of the cognition is an actual existential synthesizing that is taking place before its gaze. In the other meaning, the object is a static representation of that synthesizing, even though that synthesizing is no longer taking place." Joseph Owens, "Judgment and Truth in Aquinas," in John Catan ed., *St. Thomas Aquinas on the Existence of God: Collected Papers of Joseph Owens, C.Ss.R.* (Albany: State University of New York Press, 1980), p. 47. Also, "Accordingly, 'judgment,' in its technical sense of knowing existence, is a different activity from the constructing of propositions." *An Interpretation of Existence* (Houston: Center for Thomistic Studies,

likewise ambiguous. On the one hand, the intellect can be composing its awareness of the abstracted universal with its awareness of the singular. On the other hand, the intellect can be composing the proposition or enunciation. As far as *De Ver.* I, 3 and 9 go, the texts seem to employ my latter mentioned senses of judgment and composition.

Different as these senses of judgment and composition are a relation exists between them. As noted, the *secunda operatio intellectus* has as its sign the enunciation, or proposition. Thanks to this relation of signification, truth comes to be in the intellect. The reason is that now the intellect has something found in itself, viz., the proposition, that nevertheless corresponds to the thing as that is the object of the cognitional activity sense of judgment.

IV

I have tried to explain what Aquinas means by saying at *De Ver.* I, 9c, that truth comes to be in the intellect insofar as the intellect's judgment is about the thing as it is. Aquinas' next point is that this truth becomes known when the intellect reflectively grasps the proportion of its act to the thing. Given my interpretation so far, Aquinas should mean that truth is known when the intellect grasps the proportion between the proposition and the object of judgment in the cognitional activity sense. I should now mention that up to this point my interpretation of *De Ver.* I, 9, differs from Charles Boyer's. In a famous commentary on *De Ver.* I, 9, Boyer makes two points. First, he identifies the intellect's reflective grasp of the relationship of its act to the thing with the intellect's act of judgment, its

1985), p. 22; "What is known *dynamically* through judgment is represented *statically* in a proposition." *Ibid.*, p. 24.

secunda operatio.[12] In short, Boyer identifies judgment with the judgment of truth. In contrast, I distinguish two other senses of judgment, or *secunda operatio intellectus*, - viz., the cognitional activity sense and the propositional sense, and I have these occurring antecedent to the truth judgment. For his first point, Boyer provides two texts. First, *S.T.* I, 16, 2c:

> When, however, [the intellect] judges that a thing corresponds to the form which it apprehends about that thing, then it first knows and expresses truth. This it does by composing and dividing; for in every proposition it either applies to, or removes from, the thing signified by the subject some form signified by the predicate.[13]

Aquinas says that the intellect first knows truth in its judgment that a thing corresponds to the form it has apprehended. Boyer takes this text to mean that knowledge of the truth constitutes judgment, the intellect's second act.

Second, Boyer cites *In VI Meta.*, lect. 4, n. 1236:

> Only in the intellect's second operation is there truth and falsity, not only insofar as the intellect has a similitude of the thing known but also insofar as it

---

[12] "The judgment is essentially an act of reflection, a return of the intellect on itself." Charles Boyer, "The Meaning of a Text of St. Thomas: *De veritate*, Q. 1, A. 9," edited as the appendix in Peter Hoenen, *Reality and Judgment according to St. Thomas* (Chicago: Henry Regnery Company, 1952), p. 297. Also, "The reflection [on truth] . . . is that which has place in every judgment and which constitutes every judgment." *Ibid.*, p. 299.

[13] Cited by Boyer, *op. cit.*, p. 298. Trans. from Anton Pegis, *The Basic Writings of St. Thomas Aquinas* (New York: Random House, 1945), I, p. 171.

> reflects upon the similitude by knowing and judging it.[14]

Here Boyer takes Aquinas to be identifying second operation, reflection, and judging in respect to knowledge of truth.

Boyer's second interpretative point is that knowledge of truth occurs in a reflection upon the content of the mind's first act of simple apprehension: ". . . judging consists in declaring the agreement of a simple apprehension with a given thing, . . . judging is the apprehending of the conformity of a simple apprehension with the thing apprehended."[15] In contrast to Boyer, I have knowledge of truth occuring in respect to the propositional sense of judgment, i.e., upon the *signum* of the cognitional activity sense. In his behalf, Boyer provides the following three reasons. First, at *De Ver*, I, 9c, Aquinas uses the same term, "*iudicium*," to denote the act preceding the reflective knowledge of truth and the act of sense. Boyer then notes,

> Now the act of the sense is certainly not a judgment properly so called. . . . The act of the intellect which corresponds to it, so much so that it is defined by the same formula, is then not a judgment in the proper sense, but the act by means of which the intellect has a representation of an intelligible as matter of a true judgment. This act is the simple apprehension.[16]

Second, Boyer cites *In VI Meta.* lect. 4, n. 1236:

> For when the intellect conceives what a mortal rational animal is, it has in itself a likeness of man; but not on that account does it know that it has this

---

[14] Boyer, *op. cit.*, p. 299. My trans.

[15] *Ibid.*, p. 301.

[16] *Ibid.*, p. 300.

> likeness, because it does not judge that man is a mortal rational animal: and this is why truth and falsity are found in this second operation of the intellect, by reason of which not only does the intellect have a likeness of the understood thing, but it reflects upon the likeness itself, knowing and judging it.[17]

Here the likeness upon which the intellect reflects to know truth is its conception of mortal rational animal. This certainly sounds like a reference to an object of the intellect's first act. Thirdly and in the same vein as the just mentioned second reason, Boyer cites once more I, 16, 2c:

> But the intellect can know its own conformity with the intelligible thing; yet it does not apprehend it by knowing of a thing *what a thing is.* When, however, it judges that a thing corresponds to the form which it apprehends about that thing, then first it knows and expresses truth.[18]

Here the intellect compares to the thing the form that the intellect apprehends about the thing. Again, this sounds like a reference to the object of the intellect's first act.

In sum, Boyer's first point is that the intellect's *secunda operatio* should be *identified* with its judgment of truth. In contrast, I distinguish two other senses of judgment, viz., the cognitional activity sense and the propositional sense, and I have these occurring prior to the truth judgment. Boyer's second point is that knowledge of truth happens through a reflection upon the content of simple apprehension. In contrast I have it occurring in respect to Aquinas' propositional sense of judgment, the *signum* of the intellect's *secunda operatio.*

---

[17] *Ibid.*, p. 301. Boyer's trans.

[18] *Loc. cit.*

Does Boyer effectively oppose my *De Ver* I, 3c, interpretation of *De Ver* I, 9? I do not think so. Regarding his first reason for point one, I, 16, 2c, fails to identify the intellect's second operation with knowledge of truth. The intellect's composition and division is described in terms of its constructing of propositions. This constructing is certainly not identical with judging truth. At best, the truth judgment involves what is described in that text as the intellect's second act; it cannot be identified with that act. *In VI Meta.*, lect. 4, n. 1236, also fails to identify the truth judgment and the intellect's second act. The text distinguishes the truth judgment from the second operation as a part from its whole. The text contains the truth judgment in what it calls the second operation. In sum, the text presents a fourth use of *secunda operatio.* Not only can the phrase bear upon the two senses of judgment that I have delineated, the phrase can also bear upon the *totum* that includes those two senses plus the knowledge of truth sense noted by Boyer.

Turning to Boyer's second point, let me begin by saying that in the light of Aquinas' absolute consideration gloss of the intellect's first act, it is highly doubtful that the truth judgment reflects upon the intellect's knowledge of the thing's essence. The first operation grasps the thing's essence,[19] Aquinas equates this with absolute consideration.[20] But earlier in the *De Ente et Essentia* Aquinas provides a strong argument for his clear assertion that in absolute consideration the nature abstracts from every existence: *Patet ergo quod natura hominis absolute considerata abstrahit a quolibet*

---

[19] *Supra*, n. 4.

[20] "Secondly, by way of a simple and absolute consideration; . . . This is what we mean by abstracting the universal from the particular, or the intelligible species from the phantasm; in other words, this is to consider the nature of the species apart from its individual principles represented by the phantasms." *S.T.* I, 85, 1c; Pegis, *Basic Writings*, I, p. 814.

esse.[21] The object of the intellect's first act, then, represents no existence and so should be unfitting matter for the truth judgment. Again, as stated by *De Ver*. I, 3, truth requires something not simply proper to the intellect but also corresponding to the thing outside the soul. How is the existentially neutral common nature a credible intelligible correspondent to an existent?

What about Boyer's three textual arguments for his second point? Do they not affirm that the subject of the truth judgment is the essence of the thing as grasped in the intellect's first act? For two reasons, I think not.

First, the only apparent reason for Boyer's claim that "judgment" is used of the senses "improperly" is his mentioned first point: viz., judgment is constituted by complete reflection. Because the senses fail to do that, then judgment properly speaking fails to exist in them. But I have already replied to Boyer's first point and given reasons to dispute the equation of judgment with the complete reflection found in knowledge of truth. In Aquinas other senses of judgment exist. One sense is the intellect's apprehension of the thing's existence, and I referred to it as the cognitional activity sense of judgment. For Aquinas, however, the senses do grasp existing things and so can be said to judge.[22] Also, my

---

[21] Aquinas' argument is: ". . . but none of these beings belongs to the nature from the first point of view, that is to say, when [the essence] is considered absolutely. It is false to say that the essence of man as such has being in this individual: if it belonged to man as man to be in this individual it would never exist outside the individual. On the other hand, if it belonged to man as man not to exist in this individual, human nature would never exist in it." Trans. by Armand Maurer, *On Being and Essence* (Toronto: Pontifical Institute of Mediaeval Studies, 1968), p. 47.

[22] On the point that the senses come under "judgment" as the term technically means an awareness or an understanding that a thing exists, see Joseph Owens, "Judgment and Truth in Aquinas," edited by Catan, *op. cit.*, p. 37. Owens notes that the sense

interpreting of *De Ver.* I, 9, through *De Ver.* I, 3, seems to be the fair thing to do. Yet, as I think I have shown, the *iudicium* of article 9 by which truth comes to be in the intellect is no reference to the intellect's first operation. Rather, it is a reference to the propositional *signum* of the intellect's second act, viz., judgment in the cognitional activity sense. Finally, *De Ver.* I, 3c, makes plain that truth exists in the intellect forming the quiddities of things and definitions *only as ordered to the composition of the proposition.*[23] Contained just in the first operation, these quiddities should lack the wherewithal to be judged true or false.

My second comment is that carefully read neither *In VI Meta.* lect. 4, n. 1236, nor *S.T.* I, 16, 2c, precisely affirms the object of the intellect's first act as the matter of the truth judgment. Rather, that object must first become part of the proposition in order to be judged as a likeness of the thing. In a preceding paragraph of *lectio* 4, Aquinas says that simple concepts do not have truth or falsity.[24] Only complex concepts do. By complex concepts he means an affirmation that combines a predicate to a subject, in short a proposition.[25] The remainder of the *lectio* should give due regard to

---

judgment fails to distinguish the existence from the thing. That distinct appreciation of the thing's existence is the prerogative of intellectual judgment.

[23] "Haec enim compositio quae implicatur, scilicet aliquod animal est insensibile, est falsa. Et sic diffinitio non dicitur vera vel falsa nisi per ordinem ad compositionem sicut et res dicitur vera per ordinem ad intellectum." Leonine ed., p. 11. See also *supra*, n. 7.

[24] "Et cum voces sint signa intellectuum, similiter dicendum est de conceptionibus intellectus. Quae enim sunt simplices, non habent veritatem neque falsitatem, sed solum illae quae sunt complexae per affirmationem vel negationem." *In VI Meta.* lect. 4, n. 1224; Cathala ed., p. 369.

[25] "Dicitur autem hic affirmatio compositio quia significat praedicatum inesse subjecto." *Ibid.*, n. 1223.

Aquinas' locating truth in the proposition. But, as *In I Sent.* d. 19, q. 5, a. 1, ad 7m, asserted, the proposition is a *signum* of the *secunda operatio intellectus.* Clearly, the truth judgment relates to the second operation and its propositional product, not the first operation. I, 16, 2c, can be taken in this same vein; i.e., the intellect judges that a thing corresponds to the form which it apprehends about that thing by ". . . composing and dividing: for in every proposition it either applies to, or removes from, the thing signified by the subject some form signified by the predicate." I can read this text as reiterating the clear point of *De Ver.* I, 3c, and *In VI Meta.* lect. 4, n. 1224, that the object of the intellect's first act can be called true only in reference to its composition in the proposition. But, again, the proposition is the *signum* of the intellect's second act.[26]

In sum, *pace* Boyer, judgment, or *secunda operatio,* is not identical with knowledge of truth, though it can be that. Also, when judgment is judgment of truth, the proposition, the *signum* of the cognitional activity sense of judgment, is the subject or matter of the truth judgment.[27]

---

[26] I, 16, 2c, does contain the line that truth is in the intellect knowing *what a thing is* as in something that is true. In light of what I have said, I believe that this line is justifiably construed with the proviso: "in relation to the proposition."

[27] Despite these differences, it is important to note that both Boyer and myself understand the truth judgment as bearing upon a mental act whose content the intellect has drawn from the sensible real. For Boyer that content is the simple essence, for me that content is the proposition as it copulatively relates the essence to the real individual represented in the subject. No constitutive *apriori* factor precedes the formation of the proposition. This point of agreement makes possible my subsequent use of Boyer to handle the transcendental interpretation of *De Ver.* I, 9.

V

After saying that the intellect knows the truth it possesses by reflecting upon its own act and the proportion of its act to the thing, *De Ver.* I, 9c, goes on to say that this proportion cannot be known without knowing the nature of the intellect itself. Aquinas summarizes this line by saying that the intellect knows truth by reflecting upon itself. What is Aquinas driving at? The answer should be clear if one recalls a noted line from article 3. There Aquinas insisted that truth is the adequation of the thing and the intellect. Hence, truth requires two things. First, truth requires something proper (*proprium*) to the intellect and not possessed by the thing outside the soul. Second, this something proper to the intellect must nevertheless correspond to the extra-mental thing so that an adequation results. As I noted, the thing proper to the intellect is the propostion. Obviously, then, knowledge of truth is impossible without self-awareness. The intellectual knower must be capable of reflecting completely on itself to apprehend the item proper to itself and corresponding to the thing outside the soul. Without this complete self-reflection, our awareness of the existence of truth cannot come to be.

That is all Aquinas is intending to say. One cannot construe it as a transcendentalist analysis intent upon uncovering an *a priori*. The reflection bears upon the intellect as it sees itself proceding *a posteriori* and producing within itself a true proposition.[28] The

---

[28] In fact, according to Boyer, *op. cit.*, pp. 306-7, the very grammar of the passage is at odds with a transcendental interpretation. Aquinas says that the proportion of the intellect's act to the thing is not able to be known unless the nature of the act has been known: *nisi cognita natura ipsius actus*. Here the past tense is employed. Yet Aquinas shifts to the present tense when talking of knowledge of the intellect's nature. The nature of the act is not able to be known unless the nature of the active principle is

Transcendental Thomist undoubtedly will object that my commentary fails to do justice to the text. For as the condition for knowing truth, Aquinas speaks of the intellect knowing its nature, not just itself. This talk of knowing the nature of the active principle indicates a more profound knowledge than the intellect's self-awareness.

In reply, two good reasons exist to take the words "unless the nature of the active principle is known" simply as "unless the intellect itself is known." In other words, the first is merely a more formal way of stating the second. First, at *In II Sent.* d. 19, q. 1, a. 1c, Aquinas argues the incorruptibility of the human soul. As a third reason for the soul's having an absolute operation, i.e., one in which no corporeal organ participates, Aquinas mentions that the intellect understands itself: *intellectus intelligit se.* The elaboration of the third reason has two striking parallels to the remainder of the *responsio* of *De Ver.* I, 9. The first parallel is that Aquinas gives Avicenna's explanation of the lack of self-reflexivity in a sense power, viz., the need for an organ as a medium would split the power in half. The second parallel is that Aquinas cites the same text of the *Liber de causis: Omnis sciens qui scit essentiam suam, est rediens ad essentiam suam reditione completa.* Noteworthy is how Aquinas brings in this quote. He introduces it by way of a summarizing remark simply about the soul's knowledge *of itself.* Aquinas' transposition of self-knowledge into the knower's knowledge of its essence seems a merely terminological concession to the author of the *Liber de causis.*[29]

---

known: *nisi cognoscatur natura principii activi.* The knowledge of the nature of the active principle seems simultaneous with knowledge of the nature of the act. This would preclude a knowledge of the intellect by transcendental method.

[29] Later in his commentary on proposition 15 of the *Liber de causis*, Aquinas says that "anima sciat essentiam suam, redeat ad essentiam suam reditione" means the same as Proculus' "omne

Second, at other places Aquinas distinguishes a twofold knowledge of the soul.[30] On the one hand, there exists by self-reflection a particular knowledge of one's own soul, its acts, and its species. On the other hand, there exists by reasoning from objects, to acts, to powers, a common knowledge of the soul. This second knowledge is a knowledge of the soul's essence and demands in the words of *S.T.* I, 87, 1c, "careful and subtle inquiry." But Aquinas already makes this division at *In III Sent.* d. 23, q. 1, a. 2, ad 3m, and there in elaborating the first kind of knowledge, he again mentions that cognitive powers using an organ as a medium cannot know the existence of their own acts. The use of this argument in the context of the soul's particular knowledge of itself indicates to me, then, that in *De Ver.* I, 9c, the context is again simply the soul's particular knowledge of itself.

In sum, *De Ver.* I, 9's talk of knowing the nature of the active principle as a condition of knowing truth understandably leads some to suspect an inchoate reference to transcendental method. But if one takes up the article in the light of article 3 and earlier texts from

---

suipsius cognitivum ad seipsum omniquaque conversivum est." *Liber de Causis et Sancti Thomae de Aquino super Liberum de Causis Expositio* (Kyoto, Japan: Institutum Sancti Thomae de Aquino, 1967), p. 90. Finally, it bears pointing out that mention of the soul's self-knowledge at *C.G.* II, 49, *Item. Nullius*, and 66, *Item*, lack the *De causis* terminology.

30 ". . . non tamen removetur quin per reflexionem quandam intellectus seipsum intelligat, et suum intelligere, et speciem qua intelligit. Suum autem intelligere intelligit dupliciter: uno modo in particulari, intelligit enim se nunc intelligere; alio modo in universali, secundum quod ratiocinatur de ipsius actus natura. Unde et intellectum et speciem intelligibilem intelligit eodem modo dupliciiter: et percipiendo se esse et habere speciem intelligibilem, quod est cognoscere in particulari; et considerando suam et speciei intelligibilis naturam, quod est cognoscere in universali. Et secundum hoc de intellectu et intelligibili tractatur in scientiis." *C.G.* II, 75, *Licet.* Also, *De Ver.* X, 8c and *S.T.* I, 87, 1c.

the *Sentences* commentary, one knows how to understand the terminology - viz., to know the nature of the active principle means just to know the active principle itself.[31]

VI

In conclusion, the intellect's knowing truth by reflecting upon itself is perfectly satisfied by thinking the claim through in the light of *De Ver.* I, 3c. Consequently, the claim means that the intellect's capacity to self-reflect grasps the proportion between the thing and the proposition about the thing formed within the intellect itself. To go further forces the passage out of context.[32] This result leaves me feeling quite like the

---

[31] No incompatibility exists between the intellectual knower's immediate self-awareness and the understanding that its nature is to be conformed to things. Boyer, *op. cit.*, pp. 307-8, says: ". . . when the faculty which sees the essence of the act is the same which produces it, then it grasps in a living unity both the fruit of its own activity and the natural direction of its own movement. The intellect in perceiving itself as actually knowing judges itself to be a faculty of knowing, just as a plant, if it could sense itself forming grapes, would know itself as a grapevine (grape producer)." Likewise, "To know that by nature its act makes it conformed to things and to know that is is a faculty of conforming itself to things, is this not to know one and the same truth?" *Ibid.*, p. 307. In sum, because of the intimacy of act to principle here, to know the act's conformity to the real is to know the principle's same conformity.

[32] The Transcendental Thomist might claim that my look at the previous context fails to note *De Ver.* I, 5c, that "the truth of the first principles by which we judge everything proceed from the truth of the divine intellect as from its exemplary cause." But see the *aposteriori* gloss of this text by Knasas, "Transcendental Thomism and the Thomistic Texts," pp. 83-5. For a comprehensive working out of an *aposteriorist* epistemology, see Joseph Owens, *Cognition: An Epistemological Inquiry* (Houston: Center for Thomistic Studies, 1992).

young boy in the fairy tale "The Emperor's Clothes." Could it be that the reigning Thomism at century's end is textually naked? Many will undoubtedly find serious contemplation of the affirmative just too reactionary. Yet massive misconstrual of texts is a charge that Transcendental Thomists have hurled against their fellow Thomists. The only recourse is not continued mutual accusation but the painstaking task of reconfronting the texts. Such was my intent with *De Veritate* I, 9.

APPENDIX

THOMISTIC PERSONALISM AND TODAY'S FAMILIES*

Dr. Mary Rousseau

This effort to honor St. Thomas is offered in the spirit of Aristotle's observation that, while we cannot make an adequate return to those with whom we have studied philosophy, piety requires us nonetheless to do what we can.[1] My goal is to counter recent suggestions that we should redefine the family so as to include such current realities as broken families, blended families, adoptive families, single-parent families, child-free families, homosexual families, and so on. I want to argue that there is a normative beauty in a family consisting of two loving opposite-sex spouses and their natural children. And so, instead of revising our definition to fit contemporary realities, we would do better to adjust those realities to fit the definition. The discussion, then, is an exercise in applied ethics.

Since beauty, for St. Thomas, is the radiance of the unity, goodness, and truth of being,[2] we shall look for these three features in this normative family. As is

---

*The 1993 Center for Thomistic Studies Aquinas Lecture.

[1] See Aristotle's *Nichomachean Ethics*, IX, 1, 1164b 2-6: "And so, it seems, should one make a return to those with whom one has studied philosophy . . .; but still it is perhaps enough, as it is with the gods and one's parents, to give them what one can." In Richard McKeon (ed.), *The Basic Works of Aristotle* (New York: Random House, 1941), pp. 1077-1078.

[2] St. Thomas' notion of beauty as a transcendental attribute of all being, notionally but not really distinct from its unity, goodness, and truth, is in Chapter IV of his *Commentary on the Divine Names of Pseudo-Dionysius*. The definition of beauty as "that which, being seen, pleases," is in *S. Th.* I-II, 27, 1, ad 3. See Umberto Eco, *The Aestethics of Thomas Aquinas*, translated by Hugh Bredin (Cambridge: Harvard University Press, 1988 reprint of 1956 edition), pp. 25-36.

always the case in Thomistic ethics, ethical assessments will be rooted in philosophical anthropology and in metaphysics. The personal nature of the members of a family will determine standards for their behavior. And the sort of being that constitutes a family will specify its unity, goodness, truth and beauty. These questions are not treated explicitly in the writings of the Angelic Doctor, but we will be guided by some of his basic principles, as well as by the development of these in the writings of the Thomist philosopher who currently sits on the throne of Peter, Pope John Paul II.[3] This Holy Father's sexual and family ethics is seen by many as a crotchety Polish litany against all the little pleasures of life. But few of his critics discuss or even understand the positive philosophy underlying those negations. That underlying positive philosophy is Thomistic Personalism.[4]

---

[3] The main works of Karol Wojtyla/Pope John Paul II containing this development are: *Love and Responsibility* (translated by H. T. Willets, New York: Farrar, Straus and Giroux, 1981); *On the Dignity and Vocation of Women* (Boston: Daughters of St. Paul, 1988); *Original Unity of Man and Woman* (Boston: Daughters of St. Paul, 1981). See also Joseph Cardinal Ratzinger, *Respect for Human Life in Its Origin and on the Dignity of Procreation* (Boston, Daughters of St. Paul, 1987). For an analysis of the philosophical structure of *The Dignity and Vocation of Women*, see this author's article in *Communio International Catholic Review*, XVI, 2 (Summer 1989), pp. 200-224.

[4]See Karol Wojtyla, "Thomistic Personalism," forthcoming from Peter Lang publishers in a volume of Wojtyla's pre-papal philosophical papers translated by Teresa Sandok. W. Norris Clarke, S.J. has several important works which draw out a similar personalism from the writings of St. Thomas. See his "Fifty Years of Metaphysical Reflection: the Universe as Journey," in *The Universe as Journey* (ed. Gerald A. McCool, S.J., New York: Fordham University Press, 1988). Also Clarke's 1993 Aquinas Lecture, *Person and Being* (Milwaukee: Marquette University Press, 1993), and "Person, Being, and St. Thomas," *Communio*

Personalism takes persons to be the highest value in the universe, enjoying the highest degree of unity, goodness, and truth, the highest degree of being. The being of each being, the act by which it is really present in the world rather than a mere possibility, is the root of all its other perfections. A being's unity, indeed, is nothing other than its act of being as realizing, and thus unifying, all its other perfections.[5] Goodness is each being's desirability for the mere fact of its existence, revealed in each being's tendency to preserve its existential unity.[6] And truth is the conformity of a being's good unity with the mind of its Creator, the Supreme Person, Giver of Being to all else, Pure Being in Himself.[7] Beauty, then, is the radiance of all these, a radiance which, when being seen, pleases.[8] What we want to see in the normative family, then, is its act of being, its unifying bond, the desirability of that unified being, and the truth of its unified goodness. We will then be able to perceive its beauty and, in the light of that beauty, make some moral assessments of various other groupings of persons that claim the title "family."

Each human person has his or her unique act of being, distinct from the existential act of all other beings. But we find the fullness of our individual being, the fullness of our unity, goodness, and truth, the fullness of our beauty, when we transcend our individuality to come into communion with each other. The communion proper to persons is a kind of self-transcendence in which, without detriment to our original ontological uniqueness, we share each other's being as well. The

---

International Catholic Review, XVIII, 3 (Spring, 1993), pp. 134-148.

[5] *S. Th.* I, 11, 1.

[6] *S. Th.* I, 5, 3.

[7] *S. Th.* I, 16, 3.

[8] See note 2 above.

word *communion* contains this paradox in its etymology: a many turned into one without ceasing to be a many.[9]

The following principles of Thomistic Personalism will guide our examination of today's families.

I. From Moral Theory

A. Human acts, those for which we bear responsibility and are liable to praise or blame, are those that we do consciously and by free choice.

B. Our consciousness and freedom, and thus our responsibility, are affected in various complex ways by fears, desires, ignorance and violence.[10]

These two principles allow us to make judgments without being judgmental, that is, to declare certain human behaviors and institutions morally wrong without assigning praise or blame to any individuals. For as human agents, we bear responsibility only for what we do freely in the complex tangles of our emotions, our ignorance, and outside pressures. While we rightly condemn those who freely choose to do evil, we must excuse those whose freedom is compromised or blocked. We must also praise and admire those who cope heroically with circumstances beyond their control. And always, we need the wisdom to know the difference.

---

[9] St. Thomas identifies the moral good as the human existential good in *S. Th.* I-II, 18, 1-4. See also this author's "Life, Love and Community," in *Communio International Catholic Review*, XV, 2 (Summer 1988), pp. 199-214; and "The Primacy of Gender," forthcoming in the *1992 Proceedings of the American Catholic Philosophical Association* (published as a Supplement to *The American Catholic Philosophical Quarterly*, formerly *The New Scholasticism*.)

[10] See *S. Th.* I-II, 106 and Ralph McInerny, *Aquinas on Human Action* (Washington, D.C., 1993) pp. 3-24 on the conditions for moral responsibility.

C. Human acts are good or evil depending on whether they move us toward or away from the fulfillment or perfection of our human nature, that normative beauty of our personal existence. Thus, our moral good is precisely our fully human existence. It is the beauty of our self-transcendence in community.

II. From Philosophical Anthropology

A. Each human person is a substantial union of matter and spirit, a body informed by an intellectual soul for the good of that soul. The two defining powers of this soul, intellect and will, are the source of the properly human acts for which we bear moral responsibility.

This principle bids us recognize sexuality--gender, passion, reproduction, and family relationships--as integral parts of our beauty as persons.

B. Our way of knowing integrates intellection with sense experience, and our choices integrate freedom with passion.[11]

III. From the Metaphysics of Love

A. Our human, thus moral, goodness is nothing other than the fullness of our being, our transcendence of our own limited being, in union with the being of other persons, ultimately with God Himself.

B. We achieve this union only by exercising the love known as *amor amicitiae*. Such love is a human act in which we will some good to some person for that person's sake.[12]

---

[11]For a well developed Thomistic anthropology, distinguishing but synthesizing philosophy and theology, see Jean Mouroux, *The Meaning of Man* (Garden City, New Jersey: Doubleday and Co., Inc., 1961), in addition to the works cited in note 4, above.

[12] See *S. Th.* I-II, 28, 1, where St. Thomas asks "Whether Union is an Effect of Love?" He answers negatively, arguing that

C. The union of persons that is constituted by *amor amicitiae* is the moral norm for all human acts. Actions which seek that union are morally right or good. Those which aim at other goals are morally evil or wrong.

These principles, then, enable us to see in a normative family a unique beauty by which we can evaluate various alternative arrangements. These alternatives will be morally deficient to the degree that they deviate from that norm.

Pope John Paul II, in both pre-papal and papal writings, has creatively developed these Thomistic principles, in a synthesis with recent Semiotics and Phenomenology.[13] He has developed a unified ethical theory whose basic principle is the Personalistic Norm: Persons are to be loved, not used. Part of his general theory is a philosophical sexual ethics whose central core is St. Thomas' distinction between two kinds of love.[14] Any love, St. Thomas argues, is essentially an act of benevolence, of willing a good to someone. Love thus

---

union is love, that *amor amicitiae* constitutes union rather than causing it. See the rich development of this theme in Robert Johann, *The Meaning of Love* (Glen Rock, New Jersey: Paulist Press, 1966). Mary Hayden's "Love, the Center of the Christian Life," in *The Catholic Woman* (ed. Ralph McInerny, San Francisco, Ignatius Press, 1991, pp. 101-124) is also important. See also her "The Primacy of Love in Aquinas' Natural Law," forthcomning in the 1992 *Proceedings of the American Catholic Philosophical Association* (published as a supplement to the *American Catholic Philosophical Quarterly*.) This Thomistic notion of love-as-union is the theme of my *Community: The Tie That Binds*, (Lanham, MD: University Press of America, 1991.

[13] John Grondelski's selected (and annoted, in English) bibliography of Wojtyla's pre-papal philosophical writings runs to 35 typed pages. It will appear as an appendix to the McGivney Lectures, 1991, *At the Center of the Human Drama: The Philosophy of Pope John Paul II*, by Kenneth L. Schmitz, forthcoming from the Catholic University of America Press.

[14] See *Love and Responsibility* (cited above in note 3).

divides into the love of the good that we wish for someone, and the love of the person for whom we wish that good, *amor concupiscientiae* and *amor amicitiae*, respectively. The latter is the primary form of love when a person is loved for his/her own sake. Such love is "loving love," love in the full sense of the term, because its object is an end, something (someone) seen as good in his or her own right. *Amor concupiscientiae* is a secondary, deficient form of love because its object is a means to the well-being of something (someone) other than itself, namely, the person for whom it is willed.

It is through loving another person as an end, wishing him some good for his sake, that we come into union with him. The unifying bond between us is the good that we wish to him for his sake, for that good comes to belong to both of us at once. It belongs to the one loved by a real possession, ontologically. But it also belongs to its giver, the one loving, psychologically or intentionally. As a lover chooses to identify his beloved's good as his good, too, the good becomes common to the two. It is the bond of their union, making them one, even as it leaves their individual identities intact.[15]

Aristotle's homely example is quite telling: when a mother stays up at night to care for a sick child, if her love is rightly ordered, she wishes the child's health for the child's sake. In thus making the child's health her end, and willing her care to the child as a means to that end, she loves with what Aristotle called friendship of the third kind. Her love brings her into union with her child because a single good--the child's health--becomes a good belonging to mother and child in common. The child possesses his health ontologically, as part of his physical being. But that health is the mother's good as well. She possesses it spiritually, intentionally, by her free choice to make it the end or goal of her caring. Thus

---

[15] Cf. *S. Th.* I-II, 26, 4 on these two kinds of love; and 28, 1 and 3 on *amor amicitiae* as both unitive and ecstatic.

mother and child come into a wonderful, mysterious but very real unity with each other.[16] This unity leaves their separate identities intact even as it unites them through their possession of a single good in common.

Karol Wojtyla/Pope John Paul II refers to this union--in various pre-papal and papal writings--as a "communion of persons." Following his terms, then, I will henceforth refer to the "communion of persons," to "self-giving love," or simply, "love," (for *amor amicitiae*,) and to "desire" or "use" for (*amor concupiscientiae*.) Thus his Personalistic Norm--that persons are to be loved, not used--comes out of the Thomistic distinction between the two kinds of love. Persons are always to be loved for their own sakes, as ends, and are never to be used or desired as means to the well-being of anything, or anyone, other than themselves. Use (desire) is most vicious when someone seeks his own well-being as his primary goal or end, and subordinates another person as a means to that end. He then values that other for his contribution to his own well-being rather than for the other's intrinsic value as a person. The Holy Father rightly calls such use an attitude of self-seeking rather than self-giving. It is, precisely, egocentric, a violation of the Personalistic Norm, and thus morally wrong or evil.

When we use other persons, we preclude our own union with them. To adapt Aristotle's example, a mother would be egocentric, even in caring for her sick child, if she were to do so for self-serving motives--to put the child in her debt, perhaps, or to win bragging rights among her friends, thus using rather than loving him. Her care would then be morally wrong, and would not establish her union with her child. In the general ethical theory of the present pope, the Personalistic

---

[16] *Nichomachean Ethics*, IX, 4, 1166a1-9. See also the fine selection of philosophical readings on friendship compiled by Michael Pakaluk, *Other Selves* (Indianapolis: Hackett Publishing Co., Inc., 1992).

Norm is the rule for all human actions, and thus for actions in our sexual and family lives. When we use our sexual partners and family members, we treat them as a means instead of ends. We thus value them as the goods which we wish to ourselves for our own sakes. We desire them rather than loving them, and treat them as things rather than persons. Our own beauty, then--the radiance of our own full being, or unity, goodness and truth--is obscured, degraded, denied. We remain in isolation rather than forming a communion of persons, and our action is evil or wrong.

In this teleological ethics, the Personalistic Norm is the constant and primary guide for human actions because persons are made for communion. In communion with each other, we exist to the full. Alone, we are, precisely as persons, non-entities. Any communion of persons, by joining the act of being of one to the act of being of another, multiplies the unity, goodness, and truth, the beauty, of its individual members. This multiplied beauty marks the normative family, whose distinguishing feature is love integrated with biological ties and natural affections.

But the beauty of any communion of persons, including that of a family, is not ours by birth. It is, indeed, a difficult achievement, the outcome of a long and arduous conversion. For we are not born lovers. We are, instead, born as complete egoists, seeking our own comfort, pleasure and advancement without any concern for others. We are, indeed, unaware that others are persons, of equal value and importance with our own fascinating selves. As babies and young children, we use others without even knowing that we do, not even suspecting that it is possible to love them instead. The arduous conversion from our innate egoism, a conversion that is absolutely necessary if we are to find our true fulfillment, goodness, and beauty as persons, has been identified by recent developmental psychologists as the mark of a psychologically healthy adult. The transition from adolescence to adulthood, from psychological immaturity to maturity, and from emotional disorder to psychological health is the transition from narcissism to intimacy, from the egocentric use of others to an altruistic concern for them.[17]

But this transition, unlike the earlier stages of psychological growth, is not an automatic result of biological growth. It requires choice--deliberation, effort, hard work. It is, indeed, a life-long process from then on. But a complete conversion from egoism to altruism is that for which we are born. In order to make it, we need at least two essential pieces of psychological equipment: a healthy self-esteem and a solid conviction that love is real, not just an abstraction. This conversion, after all, requires us to stake our one and only life on self-giving love as our happiness and fulfillment. It means taking the Personalistic Norm as a guide for every decision we make. For those to whom love is a mere abstraction, an adolescent fantasy, a bit of wishful thinking rather than thoughtful wishing, it makes no sense to sacrifice pleasure, convenience, even time or money for the well-being of others. If we lack a conviction of the reality of love, our every choice will seek some benefit for our selves. Every choice, that is, will be egocentric. Thus the only persons who can give their very selves in love, who are psychologically capable of loving others instead of using them, are those who see love as a reality.

But a healthy self-esteem, based on one's authentic worth, is equally pre-requisite to the ability to love. Contrary to what we might expect, such self-esteem is not a foundation for egoism. Anxious doubt about our self-worth is what rivets our attention on ourselves and inclines us to use others in a constant search for reassurance. With that fixation, the first step

---

[17] Erik Erikson has a justly famous view of this development in *Childhood and Society* (New York: W. W. Norton, 1985. Second Edition).

toward loving--seeing another's well-being as worth of our devotion--cannot happen. Self-esteem, then, is a necessary condition for self-forgetfulness, for the ability to decenter our attention from our selves to the needs and the beauty of those whom we would love. Without self-esteem, the idea of loving others for their sakes does not even occur to us as a possibility, let alone as the urgent project of our lives.

The Holy Father's philosophy of the family, then, is based on the principle that persons are to be loved, not used, and bids us recognize that children are persons.[18] But the reverse is also true: persons, all of us persons, are at first children. If we are ever to move out of the egoism of childhood into adult self-giving love, our experience must somehow provide us with self-esteem and with the conviction that love is real. For we are not born with these crucial prerequisites to our conversion. We need, for twenty years and more, a day-in, day-out experience of being loved and cherished for the mere fact of our existence, and for being who we are. The normative family is the primary seed-bed of this experience. When those who know us best--our familiars--see us as good in our very being, and tell us that we are, we are inclined to believe them. And when those who see our goodness also make it the center of their steady concern and their constant joy, the reality of love gradually seeps into our conviction. Eventually we are able to go and do likewise, esteeming others and wishing them well for the mere fact of their being and being who they are. Then, and only then, can we begin the conversion from self-seeking to self-giving that will consume the rest of our lives. Without such childhood experience of being loved and not used, we might be trapped forever in our original egoism.

These natural ties of blood and emotions sort well with our matter-spirit make-up, with the integration

---

[18] See *The Role of the Christian Family in the Modern World*, sec. 14, p. 27 and *passim*.

of intellect and sense in our knowing, and of will and passion in our loving. Like the rest of our knowledge, our knowledge of love originates in our experience of love, and is developed in repeated existential judgments. These judgments find their "reality test" in further experiences. Parental love, then, is neither mere emotion, nor passionless willing, but a virtuous integration of passion and choice. Natural ties thus give parents a unique aptitude for loving rather than using their children. No one is better positioned to love a child for the mere fact of his existence than those whose passionate communion of persons is the source of that existence. It is difficult, indeed, to be harsh with a child if, every time you notice him, you are joyously reminded of how he came to be here in the first place. Our childhood home, then, when it is a communion of these persons whose love is the source of our being, is the natural seed-bed of our own conversion.

But the roots of parents' love lie deeper, in their sexual intimacy as spouses. In his sexual ethics, which is also a specific application of the Personalistic Norm, Pope John Paul II gives a higher value to human sexuality, and to the sex act in particular, than anyone else has done before.[19] He sees in that act the beauty of an especially apt instrument for building the self-esteem of spouses as well as their deep, solid, and steady belief in the reality of human love. This special power of the sex act lies in the role of our senses, and of the sense of touch in particular, in our existential judgments. Touch, of all our senses, gives us the most direct contact with reality, with the material singulars of our world. Existential judgments that terminate in this most direct of our senses are especially convincing. And sexual intercourse is an especially intense actuation of touch.

---

[19] See, for example, his recommendation that spouses deliberately cultivate sexual pleasure, even simultaneous orgasm, in *Love and Responsibility*, in the section entitled, "Marriage and Marital Intercourse, pp. 270-278.

When we experience love in this intensely tactual experience, we know love to be real in a existential judgment that is especially clear and certain. And when we receive such love over and over again, as we experience repeated sexual "success" in that loving context, self-esteem and self-forgetfulness are reinforced by passion. Regular exercise of sexual love thus builds a habit, the habit of charity that informs all virtues.[20]

For Pope John Paul II, the beauty of the act of sexual intercourse lies in its being an instrument or medium of communication--not something that we do, but something that we say to each other. Communications media are all the various means that bring persons into communion by symbolizing both self-giving love and some intellectual content or subject-matter. They are the means by which an idea in one person's mind is duplicated in the mind of another, so that both think the same thought at the same time, while knowing that they do so. Such communication is essential to any communion of persons.

A simple example will illustrate. No one can know, by any direct intuition or insight, what I think of today's weather. My opinion is my secret unless I choose to communicate it, to make it common to, someone else. Moreover, I have no direct pipeline by which I might insert my idea into another mind. I have only the clumsy and indirect method of creating a material symbol--a sound, a touch, a visual image--which someone else might hear, or feel, or see. That other's perception of my symbol is then the source from which he might draw the idea that I have put into the symbol, thus producing in his mind a more or less accurate duplicate of the idea in mine. A simple

---

[20] See this author's "Pope John Paul II's Teaching on Women," in *The Catholic Woman*, cited above in note 12. The poem at the end, "Making Breakfast," by Mary Stewart Hammond, is a fine expression of the power of sexual touching as an instrument of the conversion from egoism to altruism.

"Beautiful day, isn't it?" is thus the stuff of human communication, with all its hazards and all its glories, its power either to build or to destroy the most precious entity in the world, a communion of persons.

But communication is an instrument of communion only when it expresses the love by which we wish good to others for their sakes rather than using them for our own. Words, gestures, touches, looks--these are the media either of isolation, frustration and moral evil, or of communion, fulfillment, and beauty, depending on whether they bespeak love or use, self-giving or self-seeking. Whatever the subject-matter of a symbol might be--the weather, the bus schedule, or a political view--the basic content of all genuine communication is love. Symbols that bespeak use rather than love do not unify persons, do not, that is, really communicate. They are media of division and isolation, barriers to any communion of persons rather than communications media.

But communication, in order to be loving (and thus to be really communication rather than its sham), must be honest in its intellectual subject-matter. Were I to lie about my opinion of the weather, my motive for doing so would invariably turn out to be self-serving. And so, the basic intent of my speaking would be false as well. Lies always express use rather than love, and so, they are never forms of communication. They are doubly false, in information and in intention. The all-time, hands-down award for dishonest communication goes, of course, to Judas, whose kiss in the Garden of Gethsemane became the epitome of lies--a gesture which seemed to say "love" in both intellectual content and intention but was false on both counts.

The double honesty required for genuine communication is the keystone of the sexual and family ethics of Pope John Paul II. He sees the sex act as a natural sign whose informational content is love rather than use. Thus the sex act, when it means what it says, is not just one honest symbol among others. It is a paradigm of honest signs, thanks to the uniquely clear and dramatic way in which it symbolizes the nuptial meaning of the body. For our philosopher-pope, every person's body, in its gender-specific anatomy, is a natural symbol of the communion of persons. Our natural physical and psychological sexual differences constitute, in and of themselves, a body-language that says, without any words or even a conscious intention, that we are persons whose beauty is complete only when we love and thus enter into communion with each other. Thus every action of every person of every time and place, if we are to be true to our hylemorphic selves, should communicate. Whatever else we say, whenever we speak about anything, we must mean love and not use, or else violate the nuptial meaning of our own bodies.[21]

But there is something special about *the* sex act. In the moment of sexual ecstasy, human love finds a symbol of complete self-abandon. For a time we lose, indeed, give up, give over to a beloved other person, all remnants of our self-awareness and self-control. Sexual ecstasy thus speaks the love of self-gift in an especially clear, dramatic and intense way. As we have seen, to begin to love anyone, in any way, we must decenter our consciousness and freedom away from ourselves and toward the other whom we would love. We must wish him *his* good for his sake, and so, we must think about that good rather than our own. We must also decenter our will, our care and concern, from our own precious well-being to that of our beloved, thus making ourselves disponible, available, to him. In our ordinary loving actions, even when we concentrate intensely, we do not entirely lose ourselves. We keep some self-awareness and some self-control. But in the moment of sexual

---

[21] The nuptial meaning of the body is the central theme of the theology of the body developed in the papal talks on *Genesis* in 1979-80. See *The Original Unity of Man and Woman*, cited above in note 3.

ecstasy, every last shred of these is lost, given over to the one we love. We momentarily lose sight of who we are, and where, and when, and what we are doing. And in losing sight of these realities, we also give up our control over them. For a moment, our gift of our very selves is as psychologically complete as it can be. Such complete self-abandon is a clear and dramatic statement of self-giving love.

For our present Holy Father, it is no accident that this moment of complete self-abandon is also the moment that moves germ cells together so that a new person may come to be. We speak this uniquely total, permanent and exclusive love to each other precisely through our reproductive systems. The sex act is thus a sign whose intellectual content includes the possibility of a child, who will be born in egoism, like the rest of us, and will need some twenty years of parental love to enable him to decide to become an adult person converted to a life of self-giving love. The sex act is thus a promise to a child, a promise to nourish his self-esteem and his conviction of the reality of love, as long as we both shall live.

The basic principle of our philosopher-pope's sexual ethics, then, is that the act of sexual intercourse must be honest. Those who enjoy it must mean what they say. And what they say is determined for them by the physical and psychological nature of their action. Those who speak that symbol to each other, whether they mean it or not, say "self-giving love that includes the possibility of nurturing a child into fully developed personhood." If their saying is to be a medium of communication, an instrument of their communion, they must also mean, or intend, what that sign says. Any sexual activity, then, which is not motivated by self-giving love, which is not spoken to one's permanent and exclusive spouse, or which deliberately precludes the production and nurture of a new candidate for the communion of persons is a lie. In honest sexual activity, the basic intent and the intellectual subject-matter coincide. When the sex act is dishonest, both are false, and the dishonesty is redoubled. Sexual sins, then, are not just one kind of deception. They are an epitome of dishonesty because the sex act is an epitome of communication. We do not condemn them out of a negative view of sexual pleasure, but from a Personalist view of human sexual pleasure that values, above everything else in nature, human love, honest communication, and the communion of persons.[22]

Pope John Paul II's Thomistic Personalist family ethics grows out of this same basic need for loving, thus honest, communication as the way to form a communion of persons. We now can see more clearly, perhaps, the beauty of the normative family, and the deficiencies of some contemporary arrangements that are promoted in our culture as candidates for the title "family." The normative family is marked through and through by honest communication, pre-eminently in the love-making of the spouses. Their sexual intercourse is truly, habitually, an act of *making* love. They repeatedly reinforce in each other the self-esteem and the belief in the reality of love that foster their coupled, life-long process of conversion. Moreover, their entire life together is marked by the same honest communication. They thus love each other, and never use each other, in the kitchen, in the living room, the yard, and the supermarket as in the bedroom. And they love their children, in all their interactions with them, with the same love. For parental love is included in the marital proclamation of sexual ecstasy. Family life is the carrying out of that promise.

As the conversion of the parents thus moves along over the years, they gradually convince their children of their self-worth and of the credibility of human love, so that those children are enabled to begin the same process when their time comes. Sex, then, is

---

[22] The sex act as a medium of communication that needs to be honest is the central thesis of *Love and Responsibility*.

not for children so much as children are for sex. That is, the purpose of sexual intercourse is not the mere biological production of offspring. Rather, the biological production of offspring is for the sake of their reaching maturity and daring to begin the same process of conversion that will have enabled their parents to nurture them. The whole point of having children is to hand on to a new generation the capacity for self-giving love, thus perpetuating the communion of persons. Father James Burtchaell, CSC, put this point well on a recent Phil Donahue show, when he said, "Children make their parents grow up much more than parents make their children grow up."

What, then, would be the ontological features of the normatively beautiful family with which we began this discussion? What would be its unity? In what would its goodness consist? What would be the content of its truth? And how would these radiate to its observers, so that they might be pleased by its very sight? Our basic ethical principle is this: to the extent that various groupings lack the unity, goodness, and truth of the normative family, they lack its beauty as well. They will then be judged morally deficient. Some apparent families--those in which self-giving love is utterly lacking--are not deserving of the title at all, no matter how close the biological and emotional bonds of its members, no matter how close their living quarters. Blood may be thicker than water, but love alone is thick enough to bind persons together in communion.

But a beautiful family will also suit the hylomorphic nature of its members. We are not purely spiritual persons, and so, the essential personal component of familial beauty must be conjoined with certain natural, physical features. Familial love will incorporate passion. Deliberately cultivated sexual passion will be at the heart of the parents' love for each other. Gender, heredity, and natural affection also have their place. Blood ties and family-feeling are powerful incentives toward love and toward the self-sacrifice that love requires. Heredity and emotion are part of the make-up of human beings, and thus part of a family that is most completely human. The children of a normative family, then,will be the natural offspring of their parents' sexual ecstasy. A normatively beautiful family is composed of two loving parents of opposite sex and their naturally born children.

The unity of such a family is hylomorphic, as is the being of each individual member. Genes and feelings reinforce the cement of self-giving love. The unity of a family, then, is the highest unity in the physical universe, for it is a hylomorphic communion in which human persons, the highest beings in the world of nature, enjoy a fuller existence, a greater degree of human unity, than does any other earthly communion. The goodness of such a communion is also the highest good in the physical world. And its truth bespeaks its unity and goodness, revealing to our minds the glory of human personal existence that is unified by self-giving love, love that is spoken in honest communication. And so, our incentive to foster communal family life is urgent, indeed. There is, quite literally, nothing better under the sun. Among human social achievements, none could be higher.

We now can look at several alternative arrangements in our contemporary culture for which some would claim the title "family." In the descending order of their conformity to the norm, let us consider single-parent families, adoptive families, families with artificially produced children, and homosexual couples raising adopted children. Obviously a grouping that lacks love and honest communication will fall short of normative familial beauty. But those lacking certain natural, physical or biological features will be deficient as well.

As for single-parent families, surely single parents can exercise self-giving love for their children (and for their absent spouse), and often do so to a heroic degree. Granted that a single parent is chaste (for sexual activity on the part of an unmarried person, or a separated married person, is always dishonest and thus unloving), such a family meets the basic requirement for our normative beauty. Indeed, it would be more of a family, more beautiful and real, than a two-parent family that would lack either love or chastity. But the most loving single-parent family has some undeniable deficiencies in comparison to a loving two-parent family. The lack of a spouse precludes the powerful emotional confirmation of the single parent's self-esteem and belief in the reality of love that would come from regular love-making. And that emotional deficiency could affect, to some degree, the development of the children as regards their self-esteem and belief in the reality of love. Their preparation for their conversion from egoism to love could well be less complete than if they had two loving parents. One person simply cannot do the loving that two might do, and a person of one sex cannot supply the complementary gender-specific love of a person of the other sex.

Because of these deficiencies, we must distinguish between single parents who are so by choice and those who are so because of circumstances beyond their control. When single parents lovingly cope with circumstances beyond their control, their heroic virtue can go a long way toward making up for the lack of a second parent for their children. Their chaste loving can be the super-glue of a genuine communion of persons. But the character of a family is quite different if its single parent has freely chosen that state. Women who deliberately have children out of wedlock because they wish to have that experience for themselves before their biological clock runs out deliberately impose a deficient family-life on their children. They would seem, on the face of it, to have self-serving motives, and thus to give children an ongoing experience of being used rather than loved. They would then reinforce the children's innate egoism, in an upbringing that would weaken rather than strengthening their self-esteem, and that would darken rather than illuminating for them the credibility of human love. So long as these single-by-choice parents continue in this way to use their children rather than loving them, the basic requisite for a family is absent. Something similar would be true of teen-agers who produce babies out of wedlock in order to gain a sense of self-esteem for themselves, status among their peers, or someone (a child) to love them and make them feel important. In all of these cases, egocentric motives are the source and the ongoing basis of the children's very present in the world.

Loving heterosexual couples who adopt children that, for some reason other than the adoptive parents' choice, do not have a loving home with their natural parents can also meet the minimum requirement for a familial communion of persons. Their love-making must be honest, and the self-giving love of their bedroom must carry over into their concern for the well-being of their children. Assuming that their motive for adopting children is not self-serving, their sexual intimacy can promote their own ongoing conversion as they gradually build each other's self-esteem and belief in the reality of love. Children in such a home, since they would be loved rather than used, could develop the self-esteem and belief in the reality of love that would empower their eventual transition from adolescent egoism into adult loving.

But the missing natural link between the children's mere existence and their adoptive parents' sexual ecstasy, the absent genetic ties, and the lack of natural parent-child feelings remain serious deficiencies. And so, an adoptive family, while it can be more beautifully communal than a natural family that is lacking in love, remains inferior to a natural loving family. It lacks the unity, the goodness, and the truth that are supplied by the physical part of the make-up of its members. Children are better off in a loving adoptive family than in a natural family that is lacking in love--better off precisely as persons needing conversion from egoism to love. But still, an adoptive family at its best remains second best to a natural family at its best. For we are creatures of nature, not just of law, or even of love.

Our other two arrangements--families with artificially produced children and practicing homosexual couples with adopted children--can be treated together as far as the minimum requirement for a genuine family is concerned. Neither of these meets the test of having the unity generated by self-giving love. Neither, then, can be a communion of persons. For these households come into being through persons using, not loving, other persons. Lacking thus in the unity of a genuine family, they lack also its goodness and truth, its beauty, its moral goodness.

But there are some important differences, as well. The Vatican's condemnation of artificial reproduction (*in vitro* fertilization, surrogate motherhood, artificial insemination, etc.) can, indeed, seem harsh at first glance.[23] Why not welcome a technology that can relieve the anguish of sterile couples? Are not their desires often valid, their motives noble? Do they not sometimes show promise of being better parents than many natural parents? Who could fail to rejoice in the mere fact of the existence of Baby Louise Brown, Baby M and the thousands of other "test tube" babies who would not otherwise exist?

The answer is not primarily in the deficiencies of the technology, which presently requires the production and subsequent disposal of "extra" embryos. Artificial reproduction is a use, an abuse, even of the children whom the technicians count as their successes. Here we see the value of human sexuality more clearly than anywhere else. For, in the family ethics of Pope John

---

[23] See Cardinal Ratzinger's *Respect for Human Life*, cited above in note 3, where the author strongly recommends that these technologies for producing children be forbidden by law (pp. 35-38).

Paul II, the artificial production of children deliberately deprives them of their best chance to acquire the self-esteem and belief in the reality of love that are pre-requisite to their conversion from egoism to altruism. This technology separates the conception of a child from the sexual ecstasy of the child's parents, even when the gametes are those of the couple in question and they raise the child in their home. For in providing and then later uniting their germ cells in a laboratory, these parents violate the two-fold truth of the human sex act. Their sexual ecstasy is not a promise to love the precise individual child who might result from that ecstasy. And so, they violate, in their intentions, the inherent self-gift of orgasm. They do not mean what their sex-act says, for they release germ cells destined to be joined according to the intentions of the laboratory technicians. Thus the producers of the children have no commitment at all to their personal development--let alone a marital commitment destined to be reinforced by regular love-making.

The issue, once again, is not a downgrading of sexual pleasure and technology as important human goods. It is a question of honest communication in the natural coincidence of sexual ecstasy with the joining of germ cells. That natural link between a child's beginning to be and the ecstatic love of its parents for each other is the foundation for the child's gradual growth toward the conversion which will someday enable him to become fully a person. In the mind of the present Hold Father, the conjunction of orgasm with conception is so great a good for children that every child has an absolute right to be produced that way, and then to be raised in the love that was promised in its conception. Parents who deliberately separate sexual ecstasy and conception, as the parents of artificially produced children do, deliberately deprive their children of what they most need in order to become fully actualized persons. And so, their intentions are immediately suspect. They (and the lawyers and technicians involved) put children in this deprived situation for ulterior, egocentric motives. And so, the children are used rather than loved. A household produced in this fashion thus lacks the basic requirement for being a familial communion.

Homosexual partnerships are, finally, so far down the scale of deviation from the family norm as to not have its beauty at all. St. Thomas makes sodomy the worst of the vices against chastity, because it departs the furthest from the intrinsic nature of the human sex act.[24] He does not refer here to the sex act in its mere biological nature, but to the sex act of humans in its full personal nature, which combines reproduction with an ecstatic love that is self-giving and committed to the personal development of both spouses and children. Thus the evil of sodomy is not just that it precludes even the outward appearance of possible reproduction, though that defect is essential to its moral evaluation. The full deviance is that this action which precludes reproduction is thereby dishonest, and by that very fact precludes a communion of persons as well. Sodomy, in the papal ethics developed out of St. Thomas' moral treatise, is a lie that its participants speak to each other in the clarity and drama of sexual release. Sodomy is thus intrinsically manipulative and doubly dishonest, a clear violation of the Personalistic Norm whose participants use each other by mutual consent. It is a form of sexual release which, by definition, cannot be loving because it cannot be honest, and thus cannot join its collaborators in a communion of persons.

But homosexual partners do not just use each other and prevent the conception of their natural children. Committed to an ongoing dishonest sex life in which they undermine each other's self-esteem and belief in the reality of love, they also deprive their adopted children of the crucial experience of being loved rather than used. No one could lovingly provide children with such an

---

[24] *S. Th.* II-II, 154,12, "Whether the Unnatural Vice is the Greatest Sin Among the Species of Lust?"

intense, clear and daily model of persons using rather than loving each other. And so, they use their children as well, deliberately undermining their self-esteem and darkening the credibility of human love for them in an especially powerful fashion. These unfortunate children, then, even if they should be treated tenderly and well cared for, receive an upbringing that reinforces their innate egoism. They are thus seriously damaged. Precisely in their development as persons, in their ability to stake their lives on the conversion from *eros* to the love communion, let us recall, is the point of their lives. It is their personal fulfillment, their happiness, and their beauty.

Such, then, is the view of the family in the Thomistic Personalism of Pope John Paul II. But the family is a cell in a larger communion of persons. Beauty, for St. Thomas, is "that which being seen, pleases."[25] Thanks to a certain psychological contagion, the beauty of normative families extends beyond them to draw others into their communion. People committed to self-giving love as their way of life, as the norm for all their human acts, extend that love to every person they meet. They follow the Personalistic Norm in all that they do and all that they say. They treat everyone they meet with the same honesty and love that build their own family communion. And so, they are concerned for everyone's self-esteem, for everyone's belief in the reality of human love. For even adult self-esteem and mature convictions about the reality of love, hard-won as these are, remain fragile throughout our lives. Our continuing conversion requires an ongoing experience of being loved rather than used. Members of loving families provide the rest of us with that crucial experience.

Pope John Paul II sees this contagion of love as no small thing. He envisions communal families as

---

[25] Cf. above, note 2.

centers of ever-widening circles of love, forming networks of interlinked familial communions of persons that will one day encompass the earth. As the beauty of these families comes to full flower, business, health care, education, politics, entertainment, international relations, and even war will be transformed. All of human life will then be marked by honesty and by the self-giving, self-sacrificing love that brings persons into communion. When we reach that blessed state, we will have what the Holy Father calls "a civilization of love."[26] Then will the family of man attain the unity, the goodness, and the truth that will render this Big Blue Marble, this Global Village, "something beautiful for God."[27]

---

[26] See *The Role of the Christian Family in the Modern World*, sections III and IV, especially IV, 1, pp. 74-82; see also the recent book by Norbert Hoffman, *Towards a Civilization of Love* (San Francisco, Ignatius Press, 1985).

[27] This happy phrase is the title of Malcolm Muggeridge's recent book about Mother Theresa (New York, Harper and Row, 1971).

CONTRIBUTORS

Denis J. M. Bradley is Associate Professor in Philosophy at Georgetown University.

Vernon J. Bourke is Professor Emeritus of Philosophy at St. Louis University and has recently republished his *Augustine's Quest for Wisdom.*

Victor B. Brezik, C.S.B., is Professor Emeritus of Philosophy at the University of St. Thomas, Houston, and founder of the Center for Thomistic Studies.

Raymond Dennehy is Professor of Philosophy at the University of San Francisco and is President of the American Maritain Association.

Mark Griesbach is Professor Emeritus of Philosophy at Marquette University and past president of the American Catholic Philosophical Association.

Robert J. Henle, S.J., is Professor Emeritus of Philosophy and Jurisprudence at St. Louis University and recent editor of *Aquinas' Treatise on Law* and author of *Theory of Knowledge.*

John F. X. Knasas is Professor of Philosophy in the Center for Thomistic Studies at the University of St. Thomas, Houston.

Armand A. Maurer, C.S.B., is Professor Emeritus at the Pontifical Institute of Mediaeval Studies, Toronto, and author of *Being and Knowing: Studies in Thomas Aquinas and Later Medieval Philosophers.*

Joseph Owens, C.Ss.R., is Professor Emeritus at the Pontifical Institute of Mediaeval Studies, Toronto, and recent author of *Cognition, An Epistemological Inquiry* and *Towards a Christian Philosophy.*

Peter A. Redpath is Professor of Philosophy at St. John's University, Staten Island, and Director and Board member of the Institute of Advanced Philosophical Research which publishes *Contemporary Philosophy.*

Mary Rousseau is Professor of Philosophy at Marquette University and past president of the American Catholic Philosophical Association.

Leo Sweeney, S.J., is Research Professor of Philosophy at Loyola University, Chicago, and is the author of *Authentic Metaphysics in an Age of Unreality* (2nd ed.) and the forthcoming *Christian Philosophy: Greek, Medieval, and Contemporary Developments.*

OTHER PUBLICATIONS OF THE CENTER FOR THOMISTIC STUDIES

Thomistic Papers I
Victor B. Brezik, C.S.B., ed.

Articles by Henry Veatch, Vernon Bourke, James Weisheipl, Victor Brezik, Anton Pegis, and Joseph Owens.

Thomistic Papers II
Leonard A. Kennedy, C.S.B., and Jack C. Marler, eds.

Articles by Laurence Shook, Desmond FitzGerald, Robert Henle, Francis Kovach, Joseph Owens, and Frederick Wilhelmsen.

Thomistic Papers III
Leonard A. Kennedy, C.S.B., ed.

Articles by Joseph Owens, Edward Synan, Benedict Ashley, Bernard Doering, and Gerry Lessard.

Thomistic Papers IV
Leonard A. Kennedy, C.S.B., ed.

Articles by Henry Veatch, Henri DuLac, Thomas D. Sullivan, Dennis Q. McInerny, Richard J. Connell, Joseph Boyle, and Thomas A. Russman.

Thomistic Papers V
Thomas A Russman, OFM.Cap., ed.

Articles by John F. X. Knasas, Armand Maurer, Leonard Kennedy, Thomas Sullivan, and Thomas Russman.

Wisdom from St. Augustine
Vernon J. Bourke

Fourteen very readable articles gathered from many periodicals.

An Interpretation of Existence
Joseph Owens, C.Ss.R.

The problem of existence, our grasp of existence, the characteristics of existence, the bestowal of existence, the meaning of existence.

An Elementary Christian Metaphysics
Joseph Owens, C.Ss.R.

Being, essence, knowledge, and the immaterial.

Cognition: An Epistemological Inquiry
Joseph Owens, C.Ss.R.

A basic introductory text for college courses in the philosophy of knowledge.

A Catalogue of Thomists, 1270-1900
Leonard A. Kennedy, C.S.B.

The names and works of over 2,000 writers with a Thomist reputation, arranged by century, country, and (where applicable) religious community.

Known From the Things That Are
Fundamental Theory of the Moral Life
Martin D. O'Keefe, S.J.

A theoretical and applied treatment of ethics, suitable as an introductory college text using a natural law orientation.

Substance and Modern Science
Richard J. Connell

Avoiding traditional philosophical terminology, this book shows how the notion of substance is valid in modern chemistry, physics, and biology.

All publications of the Center should be ordered from:

The University of Notre Dame Press
Notre Dame, Indiana 46556